EXPRESS REVIEW GUIDES

Math Word Problems

LEARNINGEXPRESS ®

New York

Copyright © 2008 LearningExpress, LLC.

Library of Congress Cataloging-in-Publication Data

Express review guides : math word problems.
    p. cm.
    ISBN 978-1-57685-650-5 3758 3332 10/68
    1. Word problems (Mathematics) 2. Mathematics—Problems, exercises,
etc. I. LearningExpress (Organization)
    QA63.E97 2008
    510.76—dc22   2008022587

Printed in the United States of America

9 8 7 6 5 4 3 2 1

ISBN: 978-1-57685-650-5

For more information or to place an order, contact LearningExpress at:
    2 Rector Street
    26th Floor
    New York, NY 10006

Or visit us at:
    www.learnatest.com

Contents

EXPRESS
REVIEW GUIDES

Math Word
Problems

Introduction

Solving a math problem that contains only numbers is usually straightforward. You're given a couple of numbers, an operation to perform, such as addition or subtraction, and are asked to find the answer. That same question, given as a word problem, can be a bit tougher. What operation should you use? How will you arrange the information to form a number sentence that will give you the correct answer?

This book will give you the strategies you need to solve all kinds of word problems. We'll break down each problem, word by word, identifying the keywords that signal which operations to use. Diagrams, tables, and pictures can all be used to solve these problems, and we've developed an eight-step process that can be used to solve any word problem.

WHY THIS BOOK?

A word problem, like any problem, can't be solved without a plan. After giving you a pretest, this book begins with the basics of word problems: keywords. These are the terms that will tell you what operations to use, on which numbers to perform them, what formulas to use, and the order in which each step must be done. In the next chapter, we build a plan for solving word problems: an eight-step process that takes keywords a step further—or a few steps further. We learn to write number sentences for each possible operation, solve those number sentences, and then choose the most reasonable answer. Finally, we check our work to be sure our answer is correct.

Next, we take a different approach to solving word problems: We draw pictures and build tables. This gives us a few options for finding the answer to a question. There's no wrong method; if you prefer pictures, you can solve problems that way, or you can stick with the eight-step process. We'll recommend which strategy is best for each type of problem.

After showing how to use the process to solve problems that require more than one step and how to avoid the pitfalls of word problems, we cover six math topics, with a chapter on each. If you have difficulty with word problems in a specific subject, you'll want to check out one or more of these chapters. Each contains a short subject review in case you need a refresher, and then an example of how to apply those skills to word problems. Many of these topics introduce new keywords, and by the time you've finished these chapters, you'll know what keywords to look for to solve a variety of word problems.

Finally, we finish up with the posttest: 50 questions to see how far you've come. If you have any trouble, go back and reread a chapter or two. The early, strategy-developing chapters build on each other, but the topic chapters do not. Feel free to skip around and review the areas that give you trouble. This book contains nearly 100 guided examples to show you exactly how to handle each type of problem, before giving you plenty of practice problems to work through. These chapters don't just leave you with solutions, but with complete explanations for every practice question.

This book is meant to help you with word problems, not to teach algebra, geometry, or any other topic from scratch. If you want to learn one of these topics in-depth, try another Express Review Guide. Once you have a solid foundation in these skills, pick this book up again to master word problems in those subjects.

MEASURING YOUR PROGRESS

Start by taking the pretest. This 50-question exam will test your knowledge of word problems and give you a good idea of your strengths and weaknesses. Although it is just a pretest, a complete answer explanation is provided for every question. Don't worry if you struggle with the pretest—it's designed to show you in which topics or areas you need some extra practice. If you got 100% of the pretest correct, you wouldn't need this book!

The posttest is also 50 questions. It is structured in the same way as the pretest: It measures your understanding of the skills covered in Chapters 1 through 11 of this book. After completing the posttest, compare the result to your score on the pretest. Did you improve overall? Did you learn to solve the types of problems that gave you headaches in the pretest? Because the posttest is so similar in format to the pretest, you can identify which skills you have mastered and which skills still need some practice. If you were unable to answer question 44 of the pretest correctly, check to see if you answered question 44 of the posttest correctly. If so, then you've learned how to solve probability word problems, and you've gained some valuable skills.

BREAKING DOWN THE CHAPTERS

You'll find a few common headings in each chapter.

➡ *Fuel for Thought:* We define every new term as it is introduced, with examples.

➡ *Practice Lap:* After a few examples of how to solve a particular type of problem, four or five practice problems follow. The answers to these practice problems, with full explanations, are found at the end of each chapter.

➡ *Inside Track:* Sometimes, there is more than one way to solve a problem. And sometimes, we just like to point out an interesting fact. Tips and extra information are found under this heading.

➡ *Caution!:* Common mistakes are exposed here, so you can avoid making them yourself.

➡ *Pace Yourself:* These activities help reinforce the skills covered in each chapter.

GETTING STARTED

You're almost ready to take the pretest. Make sure you have a pen or pencil and plenty of scratch paper. Although every problem can be solved without a calculator, have one on hand to check your answers. Remember, even if you find the pretest tough, you've got 11 chapters coming that will explain it all. Turn the page and get started!

PRETEST

The following exam will test your knowledge of various types of word problems. The 50 questions are presented in the same order that the topics they cover are presented in this book. If you are more comfortable with some topics than others, such as probability or geometry, you may find those questions easier than others.

The pretest will show you what kinds of word problems you know how to solve, and it will also reveal the kinds of word problems that you need to review or learn more about. If you struggle with or are unable to solve any of these questions, review the answer explanations that follow the test. The explanations also provide the chapter in which the skill being tested is taught, in case you want to skip to that chapter immediately to understand that topic fully. The answer explanations describe one way in which to solve each problem. The chapters that follow will show you a few methods for attacking word problems, including one formal process that will be used throughout the book.

Save your answers after you have completed the pretest. When you have finished learning the concepts in this book, take the posttest. Compare the results of your pretest to those of your posttest to see how you've improved.

Answer the following questions. Reduce all fractions to their simplest form.

1. What is the sum of 45 and 96?

2. Liz collects 23 seashells at the beach. If Tanya collects 31 seashells, how many more seashells will she have than Liz has?

3. Jeremy has 243 letters to mail. If he splits them evenly into nine piles, how many letters are in each pile?

4. There are 12 students on the academic decathlon team, and every member studies for five hours before their next competition. How many hours did the team study?

5. Marianna volunteers at a soup kitchen twice each week. How many times does she volunteer in a year?

6. Mount Everest is 8,850 meters high, and Mount McKinley is 6,194 meters high. If the elevation of Kilimanjaro is 5,895 meters, how much lower is it than Mount McKinley?

7. Leena bowled three games and had a high score of 171, while Christine bowled four games and had a high score of 132. What is the difference between Leena's high score and Christine's high score?

8. There are eight canoes tied up at the dock. If four people fit in each canoe, how many people can go canoeing at once?

9. A pet store has 52 goldfish stored in tanks. If there are 13 fish in each tank, how many tanks does the store have?

10. Robin, Chad, Amanda, Jim, and Chip run a race. Robin finishes ahead of Chad, but behind Amanda. Chip finishes ahead of only Chad, while Jim finishes ahead of Robin. If Amanda placed first, who placed second?

11. Evelyn sells 3-gallon jugs of water and 5-gallon jugs of water. If Brad buys eight jugs and 32 gallons of water, how many of each size jug did he buy?

12. There are 78 students at the Language Center. Nine students take just French and Italian, nine of them take just Spanish and Italian, and seven of them take just Spanish and French. If there are 22 students taking only Spanish, 12 students taking only French, and 16 students taking only Italian, how many students are taking all three languages?

13. If Rocky has 13 fewer than 75 stickers, how many stickers does he have?

14. Charlie has 32 math problems to do for homework. If he has eight fewer problems than Amita, how many problems does Amita have for homework?

15. Navi divides shirts into five categories with 13 shirts in each category. How many shirts has he organized?

16. Becky takes 41 photos. She discards six of them and arranges the rest evenly among five frames. How many photos are in each frame?

17. Meredith buys a cap for $16.95 and a scarf for $8.99. She gives the cashier a $20 bill and a $10 bill. How much change will Meredith receive?

18. A hospital contains ten floors. Every floor has 16 rooms, and every room has two patients. If one nurse is needed for every eight patients, how many nurses work in the hospital?

19. Find the product of seven and a number that is four fewer than 20.

20. What is the quotient of 18 more than six and three times two?

21. What algebraic expression is equivalent to four more than negative eight times a number?

22. Write an algebraic expression that is equal to nine fewer than the square of a number.

23. If the difference between twice a number and seven is fifteen, what is the number?

24. Eight more than negative five times a number is equal to ten fewer than the number. Find the number.

25. Four times a number is greater than six fewer than the number. Find the set of values that describes the number.

26. Sherry uses $\frac{5}{4}$ cups of bleach on a large load of laundry. If she does six loads of laundry, how many cups of bleach does she use?

27. Laura spends one-third of her allowance on clothes and one-fifth of her allowance on jewelry. If she saves the rest of her allowance, what fraction of her allowance does she save?

28. Manny must balance two scales. If there are 4.506 pounds on the first scale and 4.05 pounds on the second scale, how much weight must be added to the second scale?

29. Kelly has 156.247 feet of ribbon. It takes 2.34 feet of ribbon to make a bow. If Kelly makes 62 bows, how many feet of ribbon will she have left?

30. A $300 guitar is on sale for 22% off. What is the price of the guitar?

31. Erin's notebook contains 150 pages. She buys a new notebook that has 18% more pages. How many pages are in her new notebook?

32. The string on Jessica's kite is 30 yards long. How long is her string in inches?

33. If Lynn earns $185.76 in interest on a principal of $1,720 over three years, what is her annual interest rate?

34. A ball, starting at rest, rolls downhill and reaches a velocity of 3 feet per second after four seconds. What was the ball's rate of acceleration?

35. Stanley has a 32-ounce salt solution that is 25% salt. How much salt must he add to the solution to make it 40% salt?

36. Herbert has 12 liters of a 24% alcohol solution. How many liters of 80% alcohol solution must be added to create a 32% alcohol solution?

37. Anne sleeps eight hours each day. What is the ratio of the number of hours she is awake to the number of hours she is asleep?

38. There are 12 chairs in the barbershop. If three of them are empty, what is the ratio of occupied chairs to empty chairs?

39. The ratio of freshmen to sophomores on the junior varsity cheerleading squad is 1:3. If there are 24 cheerleaders on the squad, how many of them are sophomores?

40. The ratio of tourists to residents on the island of St. James is 4:5. If there are 624 tourists on the island, how many residents are on the island?

41. There are 854 spectators at a high school basketball game. If the ratio of home fans to visiting fans is 6:1, how many fans of the visiting team are at the game?

42. Piper's box of crayons contains six red crayons, eight blue crayons, three orange crayons, four green crayons, five yellow crayons, and four purple crayons. On average, how many of each color crayon does she have?

43. East Dune Beach has 15 lifeguards on staff. Their ages, in years, are 18, 23, 24, 22, 19, 23, 29, 33, 24, 23, 33, 24, 35, 31, and 23. What is the most common age of a lifeguard at East Dune Beach?

44. Peggy has a deck of 20 cards, numbered one through 20. If she selects a card from the deck at random, what are the chances that the number on the card will be less than nine?

45. A vending machine contains 24 bottles of regular soda, 15 bottles of diet soda, 20 bottles of fruit punch, and six bottles of iced tea. If a bottle is selected at random from the machine, what is the probability that it will be either diet soda or fruit punch?

46. Triangle *ABC* has base angles that each measure 30°. What kind of triangle is triangle *ABC*?

47. If a rectangle has a length of 15 inches and a width of 3 inches, what is the perimeter of the rectangle?

48. The area of a trapezoid is 84.6 square feet. If the length of one base is 7 feet and the length of the other is 11 feet, what is the height of the trapezoid?

49. If the circumference of a circle is 54π centimeters, what is the radius of the circle?

50. Find the volume of a sphere whose radius is 12 millimeters.

ANSWERS

1. The keyword *sum* signals addition. Add 45 and 96: 45 + 96 = 141. For more on this concept, see Chapter 1.

2. The keywords *more than* signal subtraction, even though the words do not appear next to each other in the problem. Tanya has more seashells than Liz has, so we must subtract the number of seashells Liz has from the number of seashells Tanya has: 31 − 23 = 8 seashells. For more on this concept, see Chapter 1.

3. The keywords *split* and *evenly* signal division. The number of letters is divided into nine piles, so we must divide 243 by 9: $\frac{243}{9}$ = 27 letters. For more on this concept, see Chapter 1.

4. The keyword *every* can signal multiplication or division. Since we are given the number of hours one student studies and we are looking for the number of hours 12 students study, we must multiply the number of hours one student studies by the number of students: 5 × 12 = 60 hours. For more on this concept, see Chapter 1.

5. The keyword *each* can signal multiplication or division. The problem also contains the hidden number 52, because there are 52 weeks in a year. We are given the number of times Marianna volunteers in one week, and we are looking for the number of times she volunteers in 52 weeks. We must multiply the number of times she volunteers in one week by the total number of weeks: 2 × 52 = 104 times. For more on this concept, see Chapter 2.

6. The keywords *shorter than* signal subtraction, even though the words do not appear next to each other in the problem. We are looking for the difference between the height of Mount McKinley and the height of Kilimanjaro, so we do not need the height of Mount Everest. Subtract the height of Kilimanjaro from the height of Mount McKinley: 6,194 – 5,895 = 299 meters. For more on this concept, see Chapter 2.

7. The keyword *difference* signals subtraction. We are looking for the difference between Leena's high score and Christine's high score, so we do not need the numbers of games each girl bowled. Subtract Christine's high score from Leena's high score: 171 – 132 = 39. For more on this concept, see Chapter 2.

8. The following diagram shows eight canoes, each holding four people:

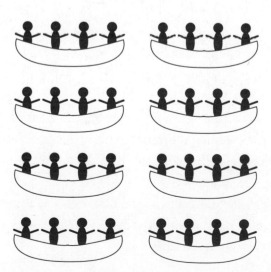

If we count the people, we can see that 32 people can canoe at once. For more on this concept, see Chapter 3.

9. The following diagram shows 52 goldfish. We can draw a circle around every 13 of them:

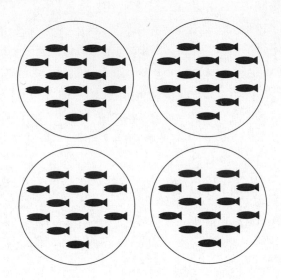

We have four circles, each with 13 fish, which means that the store has four tanks. For more on this concept, see Chapter 3.

10. We can enter these names into a table to find the order each racer finished. Robin finishes ahead of Chad, but behind Amanda. Enter Amanda at the top of the table, followed by Robin, and then Chad:

| Amanda |
| Robin |
| Chad |

Chip finishes ahead of only Chad, so place him between Robin and Chad:

| Amanda |
| Robin |
| Chip |
| Chad |

Jim finishes ahead of Robin, so we would place him on the same line as Amanda, but we are also told that Amanda placed first, so place Jim between Amanda and Robin:

Amanda
Jim
Robin
Chip
Chad

Looking at the table, we can see that Jim placed second. For more on this concept, see Chapter 3.

11. Build a table with columns for the number of 3-gallon jugs, the number of 5-gallon jugs, the total gallons from 3-gallon jugs, the total gallons from 5-gallon jugs, and the overall total. The overall total is equal to the number of 3-gallon jugs multiplied by 3 plus the number of 5-gallon jugs multiplied by 5, and the total number of jugs should always equal 8, since Brad bought eight jugs:

Number of 3-Gallon Jugs	Number of 5-Gallon Jugs	Total from 3-Gallon Jugs	Total from 5-Gallon Jugs	Total
8	0	24	0	24
7	1	21	5	26
6	2	18	10	28
5	3	15	15	30
4	4	12	20	32

If Brad buys eight jugs and 32 gallons of water, then he bought four 3-gallon jugs and four 5-gallon jugs. For more on this concept, see Chapter 3.

12. Draw a Venn diagram. We must draw it as shown on the next page, so that there is an area where just Spanish and French overlap, an area where just Spanish and Italian overlap, an area where just French and Italian overlap, and an area where all three overlap. We are given the numbers of students who take two languages and the number of students who take only one language, so we can label those areas of the diagram. The

unknown area, the number of students taking all three languages, is labeled with a variable:

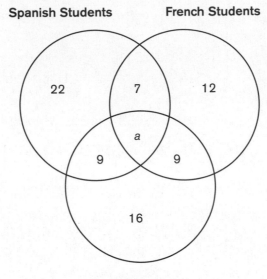

There are 78 students at the Language Center, which means that $22 + 7 + 9 + a + 9 + 16 + 12 = 78$, $75 + a = 78$, and a, the number of students taking all three languages, is 3. For more on this concept, see Chapter 3.

13. The keywords *fewer than* signal subtraction, although we must be careful to put the operands in the right order: $75 - 13 = 62$ stickers. For more on this concept, see Chapter 4.

14. The keywords *fewer than* in this problem are misleading. We are given the number of math problems Charlie has, and told that this number is eight fewer than the number of problems Amita has. Since Amita has eight more problems than Charlie, we must add eight to the number of problems Charlie has: $32 + 8 = 40$ math problems. For more on this concept, see Chapter 4.

15. The keyword *divides* signals division, but this operation has already occurred—Navi has already divided the shirts into five categories. Since each category has 13 shirts, we must multiply to find the total number of shirts: $5 \times 13 = 65$ shirts. For more on this concept, see Chapter 4.

16. The word *discards* signals subtraction, since the word means "to get rid of." Becky discards six of her 41 photos, so begin by subtracting 6 from

41: $41 - 6 = 35$. Then, she arranges them evenly among five frames. The word *evenly* signals division, because it describes how Becky separated, or divided, the photos. If she arranged 35 photos evenly in five frames, then there are $35 \div 5 = 7$ photos in each frame. For more on this concept, see Chapter 5.

17. First, find the total of Meredith's purchases. Add the cost of the cap to the cost of the scarf: $\$16.95 + \$8.99 = \$25.94$. Next, determine how much Becky gave the cashier: $\$20 + \$10 = \$30$. Since Becky gave the cashier $\$30$ and the items cost $\$25.94$, her change will be the difference between $\$30$ and $\$25.94$: $\$30.00 - \$25.94 = \$4.06$. For more on this concept, see Chapter 5.

18. The keyword *every* can signal multiplication or division, and that keyword appears three times in this problem. We are given the number of rooms on one floor and the number of floors, which means that we must multiply to find the total number of rooms: $10 \times 16 = 160$ rooms. We are given the number of patients in one room, so we must multiply to find the number of patients in 160 rooms: $160 \times 2 = 320$ patients. We are told that one nurse is needed for every eight patients, so to find how many nurses are needed for 320 patients, we must divide 320 by 8: $320 \div 8 = 40$. Forty nurses work at the hospital. For more on this concept, see Chapter 5.

19. We are looking for the product of seven and another number. That number is four less than 20. The keywords *fewer than* signal subtraction. The number sentence for "four fewer than 20" is $20 - 4 = 16$. The keyword *product* signals multiplication. The product of 7 and 16 is $(7)(16) = 112$. For more on this concept, see Chapter 5.

20. We are looking for the quotient of two numbers. The first number is 18 more than six. The keywords *more than* signal addition. The number sentence for "18 more than six" is $18 + 6 = 24$. The second number is three times two. The keyword *times* signals multiplication. The number sentence for "three times two" is $3 \times 2 = 6$. So, we are looking for the quotient of 24 and six. The keyword *quotient* signals division. The number sentence for "the quotient of 24 and six" is $24 \div 6 = 4$. For more on this concept, see Chapter 5.

21. Write *four* as "4," *negative eight* as "−8," and *a number* as "x." *Times* signals multiplication, so −8 times x is $-8x$. *More than* signals addition, so 4 more than $-8x$ is $-8x + 4$. For more on this concept, see Chapter 6.

22. Write *nine* as "9" and *a number* as "x." The square of x is x^2. *Fewer than* signals subtraction, so 9 fewer than x^2 is $x^2 - 9$. For more on this concept, see Chapter 6.

23. Write *twice* as "2," write *a number* as "x," write *seven* as "7," and write *fifteen* as "15." Replace *is* with an equals sign: the difference between $2x$ and 7 = 15. *Difference* signals subtraction: the difference between $2x$ and 7 = 15 becomes $2x - 7 = 15$. Add 7 to both sides of the equation, and then divide both sides by 2: $2x - 7 = 15$, $2x - 7 + 7 = 15 + 7$, $2x = 22$, $x = 11$. For more on this concept, see Chapter 6.

24. Write *eight* as "8," write *negative five* as "–5," write *a number* as "x," and write *ten* as "10." Replace *is equal to* with an equals sign: 8 more than –5 times x = 10 fewer than x. *More than* signals addition, *times* signals multiplication, and *fewer than* signals subtraction: 8 more than –5 times x = 10 fewer than x becomes $-5x + 8 = x - 10$. Subtract x and 8 from both sides of the equation: $-5x + 8 = x - 10$, $-5x - x + 8 = x - x - 10$; $-6x + 8 - 8 = -10 - 8$; $-6x = -18$. Divide both sides of the equation by –6: $-6x = -18$, $x = 3$. For more on this concept, see Chapter 6.

25. Write *four* as "4," write *a number* as "x," and write *six* as "6." Replace *is greater than* with the greater than symbol: 4 times x > 6 fewer than x. *Times* signals multiplication and *fewer than* signals subtraction, so we now have: $4x > x - 6$. Subtract x from both sides of the inequality and divide both sides by 3: $4x > x - 6$, $4x - x > x - x - 6$, $3x > -6$, $x > -2$. For more on this concept, see Chapter 6.

26. If one load of laundry requires $\frac{5}{4}$ cups of bleach, then six loads would require six times that amount: $(\frac{5}{4})(6) = \frac{30}{4} = 7\frac{2}{4} = 7\frac{1}{2}$ cups. For more on this concept, see Chapter 7.

27. Laura's total allowance is 1 whole. We must subtract $\frac{1}{3}$ and $\frac{1}{5}$ from 1. The least common multiple of 3 and 5 is 15, so convert all fractions to fifteenths. $\frac{1}{3} = \frac{5}{15}, \frac{1}{5} = \frac{3}{15}$, and $1 = \frac{15}{15}$. $\frac{15}{15} - \frac{5}{15} - \frac{3}{15} = \frac{7}{15}$. For more on this concept, see Chapter 7.

28. The weights must be equal, so to find how much weight must be added to the second scale, subtract the weight of the second scale from the weight of the first scale: $4.506 - 4.05 = 0.456$ pounds. For more on this concept, see Chapter 7.

29. If it takes 2.34 feet of ribbon to make one bow, then it will take 62 times that amount to make 62 bows. Multiply to find the total number of feet of ribbon Kelly uses to make the bows: 2.34 × 62 = 145.08 feet. Subtract this amount from the amount of ribbon Kelly has: 156.247 − 145.08 = 11.167 feet. For more on this concept, see Chapter 7.

30. If the price of the guitar is decreased by 22%, we must find 22% of $300 and subtract that amount from $300. To find 22% of $300, convert 22% to a decimal and multiply it by $300: 22% = 0.22, $300 × 0.22 = $66. The price of the guitar has decreased by $66. Subtract $66 from $300: $300 − $66 = $234. For more on this concept, see Chapter 7.

31. The number of pages in Erin's notebook has increased by 18%. To find percent increase, multiply the original value by 1 plus the percent (expressed as a decimal). Multiply the number of pages in her old notebook by 1.18: 150 × 1.18 = 177 pages. Erin's new notebook contains 177 pages. For more on this concept, see Chapter 7.

32. There are 3 feet in 1 yard. Multiply the number of yards, 30, by the number of feet in a yard, 3: 30 × 3 = 90 feet. Jessica's string is 90 feet long. There are 12 inches in 1 foot, so we must multiply the number of feet by 12 inches: 90 × 12 = 1,080 inches. Jessica's kite string is 1,080 inches long. For more on this concept, see Chapter 8.

33. The formula for interest is $I = prt$, where I, interest, is equal to the product of p, principal, r, rate, and t, time. We can rewrite the formula to solve for r, rate, by dividing both sides of the equation by pt: $r = \frac{I}{pt}$. Divide the interest, $185.76, by the product of the principal, $1,720, and the number of years, 3: $\frac{(\$185.76)}{(\$1,720)(3)} = \frac{\$185.76}{5,160} = 0.036 = 3.6\%$. Lynn's annual interest rate is 3.6%. For more on this concept, see Chapter 7.

34. The formula for velocity is $V = at$, where V, velocity, is equal to the product of a, acceleration, and t, time. We can rewrite the formula to solve for a, the rate of acceleration, by dividing both sides of the equation by t: $a = \frac{V}{t}$. Divide the velocity by the time: 3 ÷ 4 = 0.75 feet per second squared. The ball's rate of acceleration was 0.75 feet per second squared. For more on this concept, see Chapter 7.

35. Make a table with three rows and three columns. Since we are looking for how much salt will be added, our table will be about the percent and quantity of salt. The original solution is 25%, or 0.25, salt. The solution is 32 ounces in total, which means that 0.25 × 32 of it is salt. We are

adding an unknown quantity of salt, 100% of which is salt. Our final solution will be 40% salt:

	Percent concentration	Total quantity (oz.)	Salt quantity (oz.)
Original solution	0.25	32	0.25 × 32
Added	1	x	1 × x
Final solution	0.40		

The total quantity of the final solution will be 32 + x, and 0.40 of it will be salt:

	Percent concentration	Total quantity (oz.)	Salt quantity (oz.)
Original solution	0.25	32	0.25 × 32
Added	1	x	1 × x
Final solution	0.40	32 + x	0.40(32 + x)

The salt quantity is the product of the percent concentration and the total quantity: 0.40(32 + x), and also the sum of the salt quantity in the original solution and the amount added: 0.25 × 32 and 1 × x. Set these two expressions equal to each other and solve for x:

$0.40(32 + x) = (0.25 \times 32) + (1 \times x)$

$12.8 + 0.4x = 8 + x$

$4.8 = 0.6x$

$x = 8$ ounces

Stanley must add 8 ounces of salt to make the solution 40% salt. For more on this concept, see Chapter 10.

36. Herbert has 12 liters of a 0.24 solution, which means that the alcohol quantity of that solution is 0.24 × 12. We are adding an unknown quantity of a 0.80 solution, which has an alcohol quantity of 0.80 × x, and our final solution will be a 0.32 alcohol solution, with an alcohol quantity of 0.32 times the total quantity, 12 + x:

	Percent concentration	Total quantity (L)	Alcohol quantity (L)
Original (24%) solution	0.24	12	0.24 × 12
Added (80%) solution	0.80	x	0.80 × x
Final (32%) solution	0.32	12 + x	0.32(12 + x)

The variable x represents how many liters of the 80% solution Herbert must add. The final alcohol quantity is equal to the alcohol quantity of the original solution plus the alcohol quantity of the added solution: 0.24×12 and $0.80 \times x$. The final alcohol quantity is also equal to the product of percent concentration of the final solution and the total quantity of that solution: $0.32(12 + x)$. Set these expressions equal to each other and solve for x:

$(0.24 \times 12) + (0.80 \times x) = 0.32(12 + x)$

$2.88 + 0.8x = 3.84 + 0.32x$

$0.48x = 0.96$

$x = 2$ liters

Herbert must add 2 liters of 80% alcohol solution to the 24% alcohol solution to create a 32% alcohol solution. For more on this concept, see Chapter 10.

37. There are 24 hours in a day, so if Anne sleeps for eight hours, then she is awake for $24 - 8 = 16$ hours. The ratio of the number of hours she is awake to the number of hours she is asleep is 16:8. The greatest common factor of 16 and 8 is 8, so we can divide both numbers by 8: 16:8 = 2:1. For more on this concept, see Chapter 8.

38. There are 12 chairs in the barbershop, so if three of them are empty, then $12 - 3 = 9$ are occupied. The ratio of occupied chairs to empty chairs is 9:3. The greatest common factor of 9 and 3 is 3, so we can divide both numbers by 3: 9:3 = 3:1. For more on this concept, see Chapter 8.

39. If the ratio of freshmen to sophomores is 1:3, then the ratio of sophomores to freshmen is 3:1, and the ratio of sophomores to total cheerleaders is 3:(1 + 3) = 3:4. Set up a proportion to compare the ratio of sophomores to total cheerleaders to the ratio of the actual number of sophomores to the actual total number of cheerleaders. Since we don't know the actual number of sophomores, we can represent that number with x: $\frac{3}{4} = \frac{x}{24}$. Cross multiply and divide: $(4)(x) = (3)(24)$, $4x = 72$, $x = 18$. Eighteen of the 24 cheerleaders are sophomores. For more on this concept, see Chapter 8.

40. The ratio of tourists to residents is 4:5, and we are given the number of tourists on the island. Set up a proportion to compare the ratio of tourists to residents to the ratio of the actual number of tourists to the actual number of residents. Since we don't know the actual number of

residents, we can represent that number with x: $\frac{4}{5} = \frac{624}{x}$. Cross multiply and divide: $(4)(x) = (5)(624)$; $4x = 3{,}120$; $x = 780$. If there are 624 tourists on the island, then there are 780 residents on the island. For more on this concept, see Chapter 8.

41. The ratio of home fans to visiting fans is 6:1, which means that the ratio of visiting fans to home fans is 1:6, and the ratio of visiting fans to total fans is 1:(6 + 1) = 1:7. Set up a proportion to compare the ratio of visiting fans to total fans to the ratio of the actual number of visiting fans to the actual total number of fans. Since we don't know the actual number of fans of the visiting team, we can represent that number with x: $\frac{1}{7} = \frac{x}{854}$. Cross multiply and divide: $(7)(x) = (1)(854)$, $7x = 854$, $x = 122$. Fans of the visiting team make up 122 of the 854 spectators. For more on this concept, see Chapter 8.

42. The keyword *average* tell us that we must find the mean of the data set. Add the number of each color of crayon and divide by the number of colors: 6 + 8 + 3 + 4 + 5 + 4 = 30 crayons. There are six different colors (red, blue, orange, green, yellow, and purple), so the average number of each color crayon that Piper has is 30 ÷ 6 = 5. For more on this concept, see Chapter 9.

43. The keywords *most common* tell us that we are looking for the mode of the data set, or the age that occurs most often. The age of 23 years occurs four times, more than any other age occurs in the set, which means that it is the mode of the set, and the most common age. For more on this concept, see Chapter 9.

44. The chances, or probability, that an event will occur is the number of outcomes that make the event true divided by the total number of outcomes. There are 20 cards in the deck and Peggy selects one card, which means there are 20 total outcomes. Of the 20 cards, eight of them contain a number less than nine (the cards numbered 1, 2, 3, 4, 5, 6, 7, and 8, respectively), which means that the chances of Peggy selecting a card that is less than nine is $\frac{8}{20}$, or $\frac{2}{5}$. For more on this concept, see Chapter 9.

45. The probability that an event will occur is the number of outcomes that make the event true divided by the total number of outcomes. There are 24 + 15 + 20 + 6 = 65 bottles in the machine, which means that the total number of outcomes is 65. There are 15 bottles of diet soda, which

means that the probability of selecting a bottle of diet soda is $\frac{15}{65}$. In the same way, there are 20 bottles of fruit punch, so the probability of selecting a bottle of fruit punch is $\frac{20}{65}$. The probability that the bottle selected is diet soda or fruit punch is the sum of these probabilities: $\frac{15}{65} + \frac{20}{65} = \frac{35}{65}$, or $\frac{7}{13}$. For more on this concept, see Chapter 9.

46. There are 180° in a triangle. We can find the measure of the third angle of triangle *ABC* by subtracting the measures of the base angles from 180: 180 − 30 − 30 = 120°. A triangle with two congruent angles is an isosceles triangle. A triangle with an angle that is greater than 90° is an obtuse triangle. Since triangle *ABC* has two congruent angles and an angle greater than 90°, it is an obtuse isosceles triangle. For more on this concept, see Chapter 11.

47. The formula for the perimeter of a rectangle is $P = 2l + 2w$, where *P* is the perimeter, *l* is the length, and *w* is the width. Plug the length and width into this formula: $P = 2(15) + 2(3) = 30 + 6 = 36$ inches. The perimeter of the rectangle is 36 inches. For more on this concept, see Chapter 11.

48. The formula for the area of a trapezoid is $A = \frac{1}{2}(b_1 + b_2)h$, where *A* is the area, b_1 and b_2 are the bases of the trapezoid, and *h* is the height of the trapezoid. We can rewrite the formula to find *h* by multiplying both sides of the equation by 2 and by dividing both sides of the equation by ($b_1 + b_2$): $h = \frac{2A}{(b_1 + b_2)}$. Plug the area and the lengths of the bases into the formula: $h = \frac{2(84.6)}{(7 + 11)} = \frac{169.2}{18} = 9.4$ feet. For more on this concept, see Chapter 11.

49. The circumference of a circle is equal to $2\pi r$, where *r* is the radius of the circle. Divide the circumference by 2π to find the radius of the circle: $54\pi \div 2\pi = 27$ centimeters. If the circumference of a circle is 54π centimeters, then the radius of the circle is 27 centimeters. For more on this concept, see Chapter 11.

50. The formula for the volume of a sphere is $V = \frac{4}{3}\pi r^3$, where *V* is the volume of the sphere and *r* is the radius. Plug the length of the radius, 12 millimeters, into the formula: $V = \frac{4}{3}\pi(12)^3 = \frac{4}{3}\pi(1{,}728) = 2{,}304\pi$ millimeters³. If a sphere has a radius of 12 millimeters, then it has a volume of $2{,}304\pi$ cubic millimeters. For more on this concept, see Chapter 11.

The Keywords of Word Problems

Now that you've completed the pretest, we're ready to start breaking down some word problems. Often, the hardest part of a word problem isn't the computation—it's figuring out what operation to use. In this chapter, we'll look at the keywords that signal addition, subtraction, multiplication, and division.

In math problems, a **keyword** is a word that signals what operation to use. A problem may contain more than one keyword, especially if the word problem requires more than one operation to solve.

ADDITION

What kinds of words signal addition? How would you describe addition to someone who had never heard of it? In order for a word problem to be about addition, the author of the word problem needs to describe addition in words, so that we'll know to use that operation. If you can think of how to describe addition in words, you'll likely use some of the same words a test maker may use to write a word problem.

An addition problem is made up of two or more addends and a sum. The **addends** are the numbers that are being added, and the **sum** is the result of adding the addends together, or the answer. In the number sentence 3 + 1 = 4, 3 and 1 are the addends, and 4 is the sum.

You could say, "Addition is combining two or more numbers together to form a total." Or you may say, "Addition is the sum of a few numbers." You may also say, "Addition is putting one quantity and another together," or "Addition is one number plus another number." Each of these sentences contains one or more keywords.

Here are those sentences again, with the keywords highlighted:

Addition is **combining** two or more numbers **together** to form a **total**.
Addition is the **sum** of a few numbers.
Addition is putting one quantity **and** another together.
Addition is one number **plus** another number.

The words *combine*, *together*, *total*, *sum*, *and*, and *plus* are all keywords that represent addition. When you thought about how to describe addition, you may have used these words or words like *altogether*, *increase*, *both*, or *more*.

Example

Heather's cat weighs 8 pounds, and Serge's cat weighs 11 pounds. How much do the two cats weigh in total?

This word problem asks us to find how much the cats weigh *in total*. *Total* is a keyword that signals addition, so we must add the weights of the two cats. 8 + 11 = 19 pounds. The two cats weigh 19 pounds in total.

SUBTRACTION

Now let's think about subtraction. A subtraction problem is made up of a minuend, a subtrahend, and a difference. The number from which we are subtracting is the **minuend**, the number being subtracted is the **subtrahend**, and the result of or the answer to the subtraction problem is the **difference**. In the number sentence 3 − 1 = 2, 3 is the minuend, 1 is the subtrahend, and 2 is the difference.

How would you describe subtraction? Here are a few possible definitions:

Subtraction is **taking away** one number from another.
Subtraction is the **difference** between two numbers.
Subtraction is one number **minus** another number.
Subtraction is the operation by which you **decrease** a number or quantity.

The keywords that signal subtraction are highlighted in the previous sentences. Your definition may have used those words or words like *left, more than, less than, fewer,* or *remain.*

Example

Heather's cat weighs 8 pounds, and Serge's cat weighs 11 pounds. What is the difference between the weight of Serge's cat and the weight of Heather's cat?

This example uses the same numbers as the last example, but the keyword *difference* tells us that, this time, we need to use subtraction. The difference between the weight of Serge's cat and the weight of Heather's cat can be found by subtracting the larger weight from the smaller weight: 11 − 8 = 3 pounds.

INSIDE TRACK

WHEN WORKING ON a word problem, underline the keywords. Some word problems can be long either because they contain information not needed to solve the problem or because they require more than one operation to solve. By underlining the keywords, we can help ourselves remember which operation(s) we need to solve the problem.

MULTIPLICATION

What keywords or phrases signal multiplication? Write two different sentences that describe multiplication.

_____

_____

You may have written a sentence or two like the following:

Multiplication is one number **times** another number.
Multiplication is the **product** of two or more numbers.
Multiplication is a **factor** times another factor.

A multiplication problem is made up of two or more factors and a product. The numbers that are being multiplied are the **factors**, and the result of the multiplication problem is the **product**. In the number sentence 4 × 5 = 20, 4 and 5 are the factors, and 20 is the product.

CAUTION!

THE KEYWORD _INCREASE_ can also signal multiplication, even though it can signal addition, too. Later, we'll look at how to tell when _increase_ means addition and when it means multiplication.

Multiplication can be a tougher operation to spot than addition or subtraction. When a quantity is used for *each* or for *every* amount of another quantity, we'll need to use multiplication.

Example

If Heather's cat sleeps for 13 hours each day, how many hours does it sleep over five days?

The problem tells us how many hours Heather's cat sleeps in one day, or *each* day. Since her cat sleeps 13 hours in one day, to find how many hours it sleeps over five days, we must multiply the number of hours it sleeps in one day by the number of days: $13 \times 5 = 65$ hours.

CAUTION!

INCREASE **ISN'T THE** only keyword that could mean addition or multiplication. That last example could have been trickier if it had read: "If Heather's cat sleeps for 13 hours each day, how many total hours does it sleep in five days?" That sentence uses the word *total*, even though we need to use multiplication and not addition to solve the problem. In this example, we had the keyword *each* to help us determine that multiplication was the right operation. Remember, multiplication is like repeated addition. 13×5 could also be written as $13 + 13 + 13 + 13 + 13$. That's why multiplication sometimes seems like a total.

DIVISION

Words like *quotient*, *share*, *percent*, *out of*, and *average* are all hints that tell us to use division. Words like *per* and *each* can signal division, but can also signal multiplication. Spotting those words can narrow the possible operations to multiplication and division, and then we must use the other words in the problem to help us choose which operation to use.

A division problem is made up of a dividend, a divisor, and a quotient. The **dividend** is the number being divided, the **divisor** is the number that divides the dividend, and the **quotient** is the result of division. In the number sentence $\frac{56}{8} = 7$, 56 is the dividend, 8 is the divisor, and 7 is the quotient.

Example

Serge bought 90 ounces of cat food. If his cat eats 6 ounces of food per day, for how many days can Serge's cat eat before Serge must buy more cat food?

The keyword *per* tells us to use either multiplication or division. Which should we use? We're given the total number of ounces of cat food and the number of ounces that are eaten each day. We need to find a number that is less than the total number of ounces, since Serge's cat eats 6 of those ounces every day. Division will give us a number that is less than 90. By dividing the total number of ounces by the number of ounces eaten each day, we can find the number of days the cat food will last: $90 \div 6 = 15$ days.

If that problem had been worded a little differently, we may have needed to use multiplication instead of division.

Example

Serge's cat eats 6 ounces of food per day. How much cat food must Serge buy to feed his cat for 12 days?

This word problem also contains the keyword *per*. However, this time, we're given the number of ounces eaten per day and the number of days. We're looking for the amount of cat food eaten for all of those days combined. We need to multiply the number of ounces eaten per day by the number of days: $6 \times 12 = 72$ ounces.

PACE YOURSELF

WRITE A WORD problem of your own without using *any* of the keywords you've read in this chapter. What operation is required to solve your word problem? How could a student know what operation to use by reading your word problem? Did you just discover a new keyword?

PRACTICE LAP

DIRECTIONS: For each problem, underline the keywords and choose the operation needed to solve the problem. Then, find the solution to the problem.

1. Find the product of 15 and 9.
2. Aimee has 19 more jelly beans than Andrea. If Aimee has 33 jelly beans, how many jelly beans does Andrea have?
3. Twenty students share 220 books evenly. How many books does each student receive?
4. The bull's-eye of Matt's dartboard is worth 45 points for every time it is hit. If Matt hits the bull's-eye 14 times, how many points will he score?
5. If 26 is increased by 37, what is the new value?
6. A group of 12 skiers pays $420 for lift tickets. What is the cost of a ticket per person?
7. A store receives a shipment of 55 snow shovels. If 38 of the shovels are sold, how many remain in stock?
8. What is the quotient of 126 and 42?
9. Maurice and Steve play hockey for the Blizzards. If Maurice scored 37 goals, and Steve scored 24 goals, how many goals did they score altogether?
10. Natila brings 75 euros to France. If she spends 59 euros, how many euros does she have left?

DECIDING BETWEEN MULTIPLICATION AND DIVISION: WHICH ANSWER MAKES MORE SENSE?

Multiplication and division word problems often sound alike. When trying to decide between multiplication and division, think about fact families. For instance, these facts are all part of the same fact family:

$$4 \times 6 = 24$$
$$6 \times 4 = 24$$

$$24 \div 4 = 6$$
$$24 \div 6 = 4$$

A **fact family** is a group of related equations that use the same numbers. A fact family usually pairs addition and subtraction equations, or multiplication and division equations. The number sentences $5 - 3 = 2$, $5 - 2 = 3$, $2 + 3 = 5$, and $3 + 2 = 5$ are all members of the same fact family.

We've seen how words like *each* and *per* can signal multiplication or division. When you see those words, write a number sentence using multiplication and number sentence using division. Then, solve the multiplication sentence and decide if the answer you find makes sense given the situation described by the word problem.

Example

Michelle gives $24 to her four children to spend at an arcade. If each child receives the same amount of money, how much does each receive?

The only keyword in this problem is *each*. We are given the numbers 24 and 4. Either we need to multiply 24 by 4, or we need to divide 24 by 4:

$$\$24 \times 4 = \$96$$
$$\$24 \div 4 = \$6$$

Given the situation in the word problem, which answer makes more sense? Michelle gave $24 to her children. If she had only one child, that child would receive all $24. Since she has more than one child, each child will receive less than $24. The answer to this problem will be less than $24, because $24 needs to be shared evenly among four children. Multiplication leads to a larger number, which doesn't make sense. We must divide to find how much each child receives. $24 ÷ 4 = $6.

Let's look at another word problem that comes from the same fact family.

Example

Six friends each volunteer for four hours at the local community center. How many hours do they volunteer in all?

Again, this word problem uses the keyword *each*. We are given the numbers 6 and 4. We could divide 6 by 4, or we could multiply 6 by 4:

$6 \div 4 = 1.5$

$6 \times 4 = 24$

Apply the same strategy: which answer makes more sense? Each friend volunteers for four hours, which means that one friend volunteers for four hours, and two friends would volunteer for more than four hours together. We are looking for an operation that leads to a number that is larger than four. Since each friend volunteers for four hours, we must multiply the number of friends by 4 to find how many hours they volunteer in all: $6 \times 4 = 24$ hours.

INSIDE TRACK

WHEN DECIDING BETWEEN multiplication and division, think about the given situation in terms of one. What is the cost for one person? What is the amount that one person has? How much will one person receive? Then, think about the quantity given in the problem. Will you use the amount that one person has to use in order to find out how much a group of many has to use? If so, you will need to multiply. If you can't tell how much one person has because you are given a total, you will need to divide.

Look again at the last two examples. The first does not tell you how much one child receives, but gives you a total and asks you to find how much one child receives. Division was the operation to use. The second example gives you how much one friend volunteers, and you can use that number to find how many hours six friends volunteer. Multiplication was the operation to use in that example.

PRACTICE LAP

DIRECTIONS: For each problem, decide whether to use division or multiplication using fact families or the situation described in the problem. Then, find the answer to the problem.

11. Every student in Mr. Barry's class must perform eight experiments in order to complete a lab assignment. If there are 27 students in his class, how many experiments will be performed?

12. The cost of a field trip to the zoo is $9 per person. If there are 54 people going on the trip, how much will the trip cost?

13. Robert sells 250 raffle tickets to his cousins. If he sells 50 raffle tickets to each cousin, how many cousins does Robert have?

14. It takes a website 400 milliseconds to load for every ad on a webpage. If a webpage contains four ads, how long will it take the website to display those ads?

15. Alex has 72 minutes to play golf. If he spends eight minutes per hole, how many holes of golf can he play?

 PACE YOURSELF

Write a multiplication word problem and a division word problem using the following fact family. How would a student know which operation to use for each of your word problems? How would you explain the correct way to solve each problem to a student?

$7 \times 8 = 56$

$8 \times 7 = 56$

$56 \div 7 = 8$

$56 \div 8 = 7$

SUMMARY

KEYWORDS CAN HELP us decide which operation to use to solve a word problem. We learned which of the four major operations each keyword signifies. In the next chapter, we'll work out a step-by-step process for solving word problems.

ANSWERS

1. The keyword *product* signals multiplication: $15 \times 9 = 135$.

2. The problem contains the keyword phrase *more than* (even though *jelly beans* is in between the words *more* and *than*). Since Aimee has 19 more jelly beans than Andrea, we must subtract 19 from the number of jelly beans Aimee has (33) to find the number of jelly beans Andrea has: $33 - 19 = 14$ jelly beans.

3. The keywords in this problem are *share* and *each*. *Each* can signal multiplication or division, but the keyword *share* always signals division. Since 20 students are sharing 220 books, each student will receive fewer than 220 books. We must divide 220 by 20 to find how many books each student will receive: $220 \div 20 = 11$ books.

4. The keyword *every* signals multiplication. If one bull's-eye hit is 45 points, then two bull's-eye hits will be more than 45 points—two times more. Multiply the value of a bull's-eye hit by the number of times Matt hits the bull's-eye: $45 \times 14 = 630$ points.

5. The keyword *increased* signals addition. To increase 25 by 37, we must add 37 to 25: $25 + 37 = 62$.

6. The keyword *per* can signal multiplication or division. In this problem, we are given a number of skiers and how much they all paid for lift tickets. The cost of a ticket per person will be less than $420, so we are looking for an operation that can be performed on $420 to make that value smaller. We must divide $420 by 12 to find the cost of a ticket per person: $\$420 \div 12 = \$35$. Each skier paid $35.

7. The keyword *remain* signals subtraction. The store had 55 snow shovels, but has fewer now that it has sold some of them. To find how many snow

shovels remain, we must subtract the number of shovels sold from the original number of shovels: 55 − 38 = 17 shovels.

8. The keyword *quotient* signals division. Divide 126 by 42: 126 ÷ 42 = 3.

9. The keyword *altogether* signals addition. Add the number of goals scored by Maurice to the number of goals scored by Steve to find how many goals they scored combined: 37 + 24 = 61 goals.

10. The keyword *left* signals subtraction. Natila's original total of 75 euros decreases after she spends 59 euros. Subtract 59 from 75: 75 − 59 = 16 euros.

11. 8 × 27 = 216 experiments and 27 ÷ 8 = $\frac{33}{8}$ experiments. Which answer makes more sense? Every student must complete eight assignments. If there were only one student in the class, then the class would need to complete eight assignments. If there were more than one student in the class, then the class would need to complete more than eight assignments. We are looking for an operation than makes eight greater. Multiply the number of assignments by the number of students in the class: 8 × 27 = 216 experiments.

12. 54 ÷ 9 = $6 and 54 × 9 = $486. Which answer makes more sense? The cost of one person is $9, which means that the cost of two people will be more than $9. We are looking for an operation that makes $9 greater. Multiply the number of people on the trip by the cost per person: 54 × 9 = $486.

13. 250 ÷ 50 = 5 tickets and 250 × 50 = 12,500 tickets. Which answer makes more sense? Robert sells 250 raffle tickets to his cousins. If he had one cousin, he would have sold all 250 tickets to that cousin. If he has more than one cousin, then he would have sold fewer than 250 tickets to each cousin. The tickets are split evenly among the number of cousins, which means that we need to divide: 250 ÷ 50 = 5 tickets.

14. 400 ÷ 4 = 100 milliseconds and 400 × 4 = 1,600 milliseconds. Which answer makes more sense? If a website takes 400 milliseconds to load one ad on a page, then it takes the website more than 400 milliseconds to load more than one ad on a page. We are looking for an operation that makes 400 greater. Multiply the time it takes one ad to load by the number of ads: 400 × 4 = 1,600 milliseconds.

15. 72 ÷ 8 = 9 holes and 72 × 8 = 576 holes. Which makes more sense? If Alex spends only one minute playing a hole, he could play 72 holes

of golf. Since he spends eight minutes per hole, he will play less than 72 holes of golf. We are given a total number of minutes, so we are looking for an operation that makes 72 smaller. Divide the total number of minutes Alex has by the amount of time it takes him to play one hole: $72 \div 8 = 9$ holes.

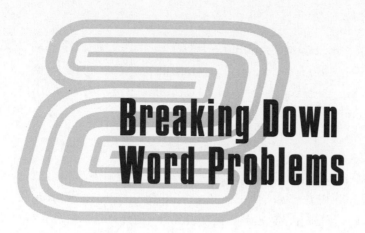

Breaking Down Word Problems

In the last chapter, we learned how to spot keywords and how to turn those keywords into mathematical operations. We also practiced writing number sentences to check the reasonableness of an operation when keywords alone weren't enough to tell us which operation to use. In this chapter, we'll look at a formal strategy for solving word problems.

What steps must we use to solve a word problem? Rather than just spell them out, let's figure them out together. How do you begin a word problem? By reading it.

STEP 1: READ THE ENTIRE WORD PROBLEM

We won't know what we need to solve until we read the problem. Some word problems can be long, so be sure to read all of the information carefully. Even if a word problem is short, be sure to read carefully. If you misread the problem, you may choose the wrong operation or set your number sentence up incorrectly. How easy is it to read too fast? Check out these three examples:

Example

Cindy's beaker contains 273 milliliters of water, and Nancy's beaker contains 237 milliliters of water. If 38 milliliters are poured from Cindy's beaker to Nancy's beaker, how much water is now in Cindy's beaker?

Example

Cindy's beaker contains 273 milliliters of water, and Nancy's beaker contains 237 milliliters of water. If 38 milliliters are poured from Cindy's beaker to Nancy's beaker, how much water is now in Nancy's beaker?

Example

Cindy's beaker contains 273 milliliters of water, and Nancy's beaker contains 237 milliliters of water. If 38 milliliters were poured from Cindy's beaker to Nancy's beaker, how much water was in Cindy's beaker?

These three word problems are very similar, but the answers to each are different. Did you catch the differences? The first problem asks us to find how much water is in Cindy's beaker after 38 milliliters are poured from her beaker into Nancy's beaker. Since water left Cindy's beaker, we need to subtract 38 from 273: 273 − 38 = 235 milliliters.

The second word problem also states that 38 milliliters are poured from Cindy's beaker into Nancy's beaker, but this problem wants us to find how much water is now in Nancy's beaker. Since the amount of water in Nancy's beaker increased, we need to add 38 to 237: 38 + 237 = 275 milliliters.

The third word problem again states that 38 milliliters are poured from Cindy's beaker into Nancy's beaker, but this problem tells us that 273 milliliters is the volume of water in Cindy's beaker after 38 milliliters were poured out, and the problem asks us to find out how much water was in Cindy's beaker before those 38 milliliters were poured out. We need to add 38 milliliters to the volume in Cindy's breaker now to find out how much water was in the beaker: 273 + 38 = 311 milliliters.

Two of these word problems required addition and one required subtraction, and even the two that used addition were different. That's why the first step in solving a word problem is to read the entire word problem, and read it carefully.

STEP 2: IDENTIFY THE QUESTION BEING ASKED

After we have carefully read the problem, it is important to identify the question being asked. In our previous three examples, we were given the same characters, the same props, and the same numbers, but the question being asked in each example was different, which made our answers different. After we have read the word problem, it is important to understand what is being asked so that we will have a better idea of what operation to be used, even after we have identified keywords.

STEP 3: UNDERLINE THE KEYWORDS

Once we have carefully read the problem and identified the question being asked, the next step is to underline the keywords. In doing so, we'll either know which operation to use, or we will have limited the possible operations to just a few. What if there are no keywords in a word problem? The three examples we looked at in step 1 did not contain any of the keywords we looked at in Chapter 1. When this happens, we must use the context of the word problem and the question we've been asked to answer to help us limit the possible operations.

FUEL FOR THOUGHT

THE *CONTEXT* OF a word problem is the situation in which the problem occurs. For instance, the numbers 3 and 5 may not mean much by themselves, but in the context of a word problem, they could be added, subtracted, multiplied, or divided. The words (including keywords) that surround those numbers are the context, and it is the context that tells us which operation we will need to use.

For example, when a word problem describes a person spending money, subtraction is often the operation to use. Spending involves a decrease, and an operation like subtraction (or division) can be used to decrease a total.

What should we do now? Having underlined our keywords and analyzed the context of the word problem, we can make a list of all the possible operations the word problem may require us to use.

STEP 4: LIST THE POSSIBLE OPERATIONS

By making a list, we can test each operation and cross out the ones that are incorrect.

Let's practice applying the first four steps to an example.

Example

Teddy walks 17 blocks to school. If Omar walks four fewer blocks to school, how many blocks does Omar walk to get to school?

1. *Read the entire word problem.*
   The problem contains the numbers 17 and four, and we are looking for the number of blocks Omar walks to get to school.
2. *Identify the question being asked.*
   We are looking for the number of blocks Omar walks. The number of blocks Teddy walks is given. We are told that Omar walks four fewer blocks than Teddy, which tells us the number of blocks Omar walks will be fewer than 17.
3. *Underline the keywords.*
   You should have underlined the word *fewer*, since we learned in Chapter 1 that this word signals subtraction.
4. *List the possible operations.*
   We already have subtraction as a possibility, but check the context of the problem.

Based on the information gathered in these steps, we are looking for an operation that will make the number 17 smaller. Using division wouldn't make sense, because we know that the keyword *fewer* means subtraction, so the only possibility is subtraction.

Some word problems won't have any keywords at all. That is why it is so important to list the possible operations that could be used to solve a word problem. You may have to list many operations, and then compare the results of those number sentences to the context of the problem to see which answer makes the most sense.

STEP 5: WRITE NUMBER SENTENCES FOR EACH OPERATION

In steps 1 through 4, we've limited the number of possible operations. What should we do next? Use those operations and the numbers given in the problem to form number sentences.

By writing number sentences for each operation, we can determine which operations make sense. We can also determine if we've placed the numbers, or operands, in the correct position in the number sentence. For some operations, like addition and multiplication, the order of the operands doesn't matter, but for operations like subtraction and division, order is very important.

INSIDE TRACK

ADDITION AND MULTIPLICATION are commutative operations, which means that the order of the addends in an addition problem and the order of the factors in a multiplication problem do not matter. 3 + 4 and 4 + 3 both equal 7, and 3 × 4 and 4 × 3 both equal 12. However, subtraction and division are not commutative. 4 − 3 = 1, but 3 − 4 = −1. 12 ÷ 3 = 4, but 3 ÷ 12 = 0.25. If you are considering using subtraction or division to solve a word problem, be sure to consider all possibilities for the order of the operands and eliminate the number sentences that do not make sense in the context of the problem.

Once we have written the possible number sentences for the word problem, we're ready to produce one or more possible answers. In the example of Omar and Teddy, we've decided to use subtraction, and the problem contains the numbers 17 and 4. There are two number sentences we can form: 17 − 4 and 4 − 17.

STEP 6: SOLVE THE NUMBER SENTENCES AND DECIDE WHICH ANSWER IS REASONABLE

Let's solve each: $17 - 4 = 13$ and $4 - 17 = -3$. Which answer is reasonable? It would be impossible to walk -3 blocks to school. We know from the context of the problem that Omar walks fewer blocks than Teddy, and 13 blocks are fewer than 17 blocks. The reasonable answer is 13.

STEP 7: CHECK YOUR WORK

Check the answer to a problem by using the inverse of the operation used to find your answer. We used subtraction to find the answer to this problem, so we will use addition to check the answer: $13 + 4 = 17$, the number of blocks Teddy walks. We have performed subtraction correctly. Now check to see that the original question was answered correctly, "How many blocks does Omar walk to school?" Remember that we can perform the operation correctly but have the incorrect answer to the problem. The three examples in step 1 showed us different ways of reading the same problem and coming up with different answers. Make sure your solution is the answer to the question being asked.

FUEL FOR THOUGHT

TWO OPERATIONS ARE inverses of each other if one operation "undoes" the other. Addition and subtraction are inverse operations. If we add 5 and 4 to find a sum of 9, we can undo that addition by subtracting 4 from 9 to get back to 5. Multiplication and division are also inverse operations. $20 \div 4 = 5$. To go from 5 back to 20, we perform the inverse of division: We multiply 5 by 4, and the product is 20.

1. A relay race is run by five athletes. If the race is 1,000 meters, and each athlete runs the same distance, how far does each athlete run?

2. A machine produces 420 products in an hour. How many products will the machine produce in seven hours?

3. Jonathan's disc can hold 1,440 bytes. The files on the disc right now take up 865 bytes. How many bytes are remaining on the disc?

4. A truck driver made 18 deliveries last week and 24 deliveries this week. How many deliveries did the driver make altogether?

5. Allie swam 55 laps every day for 15 days. How many laps did she swim over that period of time?

6. Chris's seashell weighs 1,382 milligrams. If Cynthia's seashell weighs 1,744 milligrams, how much do the two seashells weigh in total?

7. A box holds 625 index cards. If Emma uses 313 cards, how many cards are left in the box?

8. Ethan has 12 rolls of nickels. Each roll holds 40 nickels. How many nickels does Ethan have?

9. David records 1,565 minutes of footage over five days. How many minutes of footage did he record each day?

10. A polar bear loses 45 pounds over the course of a year. If the weight of the polar bear was 903 pounds at the start of the year, what is the weight of the bear now?

We now have a seven-step process for solving word problems. Let's test our process on another example.

Example

Judah read five books in the month of May. The first book contains 241 pages, and the second book contains 312 pages. How many pages were the first two books combined?

Work through solving this problem step by step:

1. *Read the entire word problem.*
   This word problem isn't long, and we can tell that we need to use the sizes of each book to find our answer.
2. *Identify the question being asked.*
   "How many pages were the first two books combined?"
3. *Underline the keywords.*
   The word *combined* is a keyword that signals addition.
4. *List the possible operations.*
   Addition is the only reasonable operation. We need to find the total number of pages in the first two books, so operations that decrease a number, such as subtraction and division, won't work. Multiplying the book sizes would give us a huge number that is much greater than the number of pages in the two books.
5. *Write number sentences for each operation.*
   241 + 312
   312 + 241

The order of the addends in an addition sentence doesn't matter, so we can eliminate the second number sentence, since it will give us the same total as the first number sentence. What about the number 5? We are also told that Judah read five books in the month of May. This number cannot help us solve the problem; in fact, it may be included in the word problem just to trick us. Numbers like this are extraneous information, so cross out the number 5. Eliminating extraneous information will make writing and choosing the correct number sentence easier. Let's add a step to our word-problem-solving process. As a rule, after underlining the keywords and before listing the possible operations, cross out any extraneous information.

Our word problem now looks like this:

~~Judah read five books in the month of May.~~ The first book contains 241 pages, and the second book contains 312 pages. How many pages were the first two books <u>combined</u>?

6. *Solve the number sentences and decide which answer is reasonable.*
   We have only one number sentence: 241 + 312 = 553 pages.
7. *Check your work.*
   Use subtraction to check addition. Subtract the total number of pages from either the number of pages in the first book or the number of pages in the second book. The total minus the number of pages in the first book should give us the number of pages in the second book (and the total minus the number of pages in the second book should give us the number of pages in the first book): 553 − 241 = 312 and 553 − 312 = 241. The number 553 answers the question, "How many pages were the first two books combined?" Our answer is correct.

- PRACTICE LAP -

DIRECTIONS: For the remaining questions, include a step to cross out any extra information and, if applicable, translate words into numbers.

11. Monica rides the subway for 35 minutes and then walks another 13 minutes to get to work. If the subway moves at an average speed of 45 miles per hour, how long does it take Monica to get to work?

12. A stamp costs 41 cents. Lisa buys 25 stamps and uses them to mail 11 packages. How much money did Lisa spend on stamps?

13. Lindsay flies 2,462 miles in five hours, while Jamie flies 719 miles in two hours. How much farther does Lindsay fly than Jamie?

14. A concert hall contains 3,450 seats. If 3,216 seats are sold for a show, and the cost is $18 per ticket, how much money did the concert hall collect?

15. Lincoln College enrolled 2,314 students last year. If the number of students increased by 239 this year, and the cost of tuition increased by $1,200, how many students are attending Lincoln College this year?

That last example led to an improvement in our word-problem-solving process. Another example may help us improve the process further.

Example

Stephen is 15 years old, and he sleeps eight hours each day. How many hours does he sleep in a week?

Using our new step 4, work through solving this problem step by step:

1. *Read the entire word problem.*
   We are given the number of hours Stephen sleeps in a day and we are looking for the number of hours he sleeps in a week.
2. *Identify the question being asked.*
   "How many hours does he sleep in a week?"
3. *Underline the keywords.*
   The word *each* is a keyword.

4. *Cross out extra information.*

   Stephen's age is not needed to find the number of hours he sleeps in a week, so cross out that number.

5. *List the possible operations.*

   The keyword *each* can signal multiplication or division.

6. *Write number sentences for each operation.*

   Now that we've crossed out the extra information (the number 15), we are left with only one number, the number of hours Stephen sleeps in a day. We are looking for the number of hours Stephen sleeps in a week. The word *week* actually represents a number. There are seven days in a week, so we are looking for the number of hours Stephen sleeps in seven days. Let's revise step 3 of the word-problem-solving process. After crossing out extra information, we must look for words that represent numbers and translate them. Now we can write our number sentences:

   8×7

   $8 \div 7$

   $7 \div 8$

7. *Solve the number sentences and decide which answer is reasonable.*

   $8 \times 7 = 56$

   $8 \div 7 \approx 1.14$

   $7 \div 8 = 0.875$

   Since Stephen sleeps eight hours in one day, the number of hours he sleeps in seven days will be greater than 8, not less than 8, so division is the wrong operation. Multiplication correctly shows us how many hours Stephen sleeps in a week: $8 \times 7 = 56$ hours.

8. *Check your work.*

   Since we used multiplication to find our answer, we must use division to check our work. The number of hours Stephen sleeps in a week divided by the number of days in a week should give us the number of hours Stephen sleeps each day: $56 \div 7 = 8$ hours.

We now have an eight-step process for solving word problems:

1. Read the entire word problem.
2. Identify the question being asked.
3. Underline the keywords.

4. Cross out extra information and translate words into numbers.
5. List the possible operations.
6. Write number sentences for each operation.
7. Solve the number sentences and decide which answer is reasonable.
8. Check your work.

INSIDE TRACK

WORD PROBLEMS THAT contain words that must be translated into numbers are often multiplication or division problems. If you translate a word into a number and there are no keywords in the word problem, write number sentences for multiplication and division and then decide which answer is most reasonable.

Let's apply the eight-step process to another example.

Example

Ariscielle buys a round-trip train ticket from New York to Boston for $168. What is the price of the ticket each way?

1. *Read the entire word problem.*
   We are given the price of a round-trip ticket.
2. *Identify the question being asked.*
   We are looking for the cost of a one-way ticket.
3. *Underline the keywords.*
   The word *each* is a keyword.
4. *Cross out extra information and translate words into numbers.*
   There is no extra information in this problem, but there is only one number. The term *round-trip* means that the ticket can be used twice, once in each direction. Translate the word *round-trip* into the number 2. The cost of the ticket, $168, represents two fares.
5. *List the possible operations.*
   The keyword *each* can signal multiplication or division.

6. *Write number sentences for each operation.*

We have one number sentence for multiplication and two for division:

168×2

$168 \div 2$

$2 \div 168$

7. *Solve the number sentences and decide which answer is reasonable.*

$168 \times 2 = 336$

$168 \div 2 = 84$

$2 \div 168 \approx 0.12$

Since the cost of two fares is $168, the cost of one fare must be less than that. Multiplication is the wrong operation because multiplication would give us more. Our division number sentences give us quotients of $84 and approximately $0.12. The answer $0.12 is much too small. The correct number sentence is $168 \div 2 = 84$. The answer, $84, is a reasonable cost for the price of the ticket each way.

8. *Check your work.*

Since we used division to find our answer, we must use multiplication to check our work. The price of the ticket each way multiplied by the number of fares in a round-trip ticket should be equal to the price of a round-trip ticket: $84 \times 2 = $168. The answer, $84, correctly answers the question, "What is the price of the ticket each way?"

PACE YOURSELF

WE MUST REMEMBER the order of operations with the acronym PEMDAS, which stands for **P**arentheses, **E**xponents, **M**ultiplication, **D**ivision, **A**ddition, and **S**ubtraction. The phrase "Please Excuse My Dear Aunt Sally" is a mnemonic device. A mnemonic device is a word or phrase that helps us remember something—and that something can be completely unrelated to the mnemonic device, as it is with "Please Excuse My Dear Aunt Sally." Can you create a mnemonic device to help yourself remember the eight steps to solving a word problem?

16. How many inches are in 9 feet?

17. A dozen books weigh 24 pounds and cost $360. What is the cost per book?

18. If Nate has 18 pairs of socks, how many individual socks does he have?

19. If four sides of a cube are painted red, how many sides are not painted red?

20. Dylan drank 4 cups of milk and 1 cup of orange juice today. How many pints of milk did Dylan drink?

SUMMARY

IN THIS CHAPTER, we developed an eight-step process for solving word problems. We used the keywords we learned in Chapter 1 as well as the context of a problem and the question being asked to help decide which operations to use. After eliminating extraneous information and translating words into numbers, we wrote number sentences and chose the most sensible answer. In Chapter 5, we'll learn another strategy for solving word problems.

ANSWERS

1. *Read the entire word problem.*

We are given the total distance of a race and the number of athletes.

Identify the question being asked.

We're looking for how far each athlete runs.

Underline the keywords.

The word *each* is a keyword.

List the possible operations.

The keyword tells us we'll need to use either multiplication or division.

Write number sentences for each operation.

Since multiplication is commutative (the order of the factors does not matter), we have only one multiplication number sentence: $5 \times 1{,}000$. We have two division number sentences: $5 \div 1{,}000$ and $1{,}000 \div 5$.

Solve the number sentences and decide which answer is reasonable.

$5 \times 1{,}000 = 5{,}000$; $5 \div 1{,}000 = 0.005$; $1{,}000 \div 5 = 200$. The total distance of the race is 1,000 meters, so it would be impossible for each athlete to run 5,000 meters. Each athlete must run fewer than 1,000 meters. The number 0.005 is very small. Since there are only five athletes, if each ran 0.005 meters, they would be unable to finish the race, since 5×0.005 is 0.025, which is much less than 1,000. It sounds reasonable that each athlete would run 200 meters, since this number is less than 1,000, but not too small.

Check your work.

Since we used division to find the answer to this problem, we must use multiplication to check our work. If each athlete runs 200 meters and there are five athletes, then $200 \times 5 = 1{,}000$, which is the total distance of the race. The correct answer is 200 meters because not only does it make sense mathematically, but it also answers the original question, how far does each athlete run.

2. *Read the entire word problem.*

We are given the number of products produced in one hour.

Identify the question being asked.

We are looking for the number of products produced in seven hours.

Underline the keywords.

There are no keywords in this word problem, so we will need to use the context of the problem to determine which operation to use.

List the possible operations.

If the machine produces 420 products in one hour, it will produce more than 420 products in more than one hour. We are looking for an operation that will make the number 420 larger. Multiplication and addition will both make the number 420 larger.

Write number sentences for each operation.

Since addition and multiplication are both commutative, we have only one number sentence for each operation: 420×7 and $420 + 7$.

Solve the number sentences and decide which answer is reasonable.

$420 \times 7 = 2{,}940$; $420 + 7 = 427$. The answer 427 is too small. If the machine produces 420 products in the first hour, it would produce 420 products in the second hour, for a total of 840 in two hours. In seven hours, it would produce much more than 427 products. Since the machine produces 420 products each hour, multiplication is the right operation to use. The answer, 2,940, is reasonable.

Check your work.

Since we used multiplication to find the answer to this problem, we must use division to check our work. If the machine produces a total of 2,940 products in seven hours, then the number of products it produces in one hour should be equal to 2,940 divided by 7: $2,940 \div 7 = 420$. The solution, 2,940, also answers the original question.

3. *Read the entire word problem.*

We are given the size of Jonathan's disc and the number of bytes used on the disc.

Identify the question being asked.

We are looking for the number of bytes that are remaining.

Underline the keywords.

The word *remaining* is a keyword.

List the possible operations.

The keyword *remaining* signals subtraction.

Write number sentences for each operation.

We have two subtraction number sentences we can write using these numbers: $1,440 - 865$ and $865 - 1,440$.

Solve the number sentences and decide which answer is reasonable.

$1,440 - 865 = 575$ and $865 - 1,440 = -575$. The number of bytes remaining on the disc won't be a negative number, since that would be impossible. Since the disc holds 1,440 bytes and 865 bytes are used, the number of free bytes is the positive difference between those two numbers. The answer, 575 bytes, is reasonable.

Check your work.

Since we used subtraction to find the answer to this problem, we must use addition to check our work. If 865 bytes are used and 575 bytes are free, the total size of the disc, 1,440, should be equal to the sum of 865 and 575: $865 + 575 = 1,440$. The answer to the original question of how many bytes remain is 575, so our answer is correct.

4. *Read the entire word problem.*

We are given the number of deliveries made in two different weeks.

Identify the question being asked.

We are looking for the total number of deliveries made in those two weeks.

Underline the keywords.

The word *altogether* is a keyword.

List the possible operations.

The keyword *altogether* signals addition.

Write number sentences for each operation.

Since addition is commutative, we have only one number sentence to write: 18 + 24.

Solve the number sentences and decide which answer is reasonable.

18 + 24 = 42 deliveries.

Check your work.

Since we used addition to find the answer to this problem, we must use subtraction to check our work. If 42 deliveries were made altogether, then the total minus the number of deliveries made last week, 18, should equal the number of deliveries made this week, 24: 42 − 18 = 24. The number 42 gives the correct answer to the original question.

5. *Read the entire word problem.*

We are given the number of laps Allie swam in one day and the number of days she swam.

Identify the question being asked.

We are looking for the number of laps she swam in 15 days.

Underline the keywords.

The word *every* is a keyword.

List the possible operations.

The keyword *every* signals multiplication, although *every*, like *each*, can sometimes signal division.

Write number sentences for each operation.

We have one multiplication sentence and two division sentences to write: 55 × 15, 55 ÷ 15, and 15 ÷ 55.

Solve the number sentences and decide which answer is reasonable.

55 × 15 = 825, 55 ÷ 15 ≈ 3.7 and 15 ÷ 55 ≈ 0.27. The decimal answers do not make sense, since Allie swims a whole number of laps every day, which means that the number of laps she would swim over 15 days would also be a whole number. Multiplication makes sense, since we are

given the number of laps she swims in one day and we are looking for the number of laps she swims for more than one day. The answer, 825, is reasonable.

Check your work.

Since we used multiplication to find the answer to this problem, we must use division to check our work. If Allie swims 825 laps in 15 days, then 825 divided by 15 should give us the number of laps Allie swims in one day: $825 \div 15 = 55$ answers the question correctly.

6. *Read the entire word problem.*

We are given the weight of two seashells.

Identify the question being asked.

We are looking for the total weight of those shells.

Underline the keywords.

The word *total* is a keyword.

List the possible operations.

The keyword *total* signals addition.

Write number sentences for each operation.

Since addition is commutative, we have just one number sentence: $1,382 + 1,744$.

Solve the number sentences and decide which answer is reasonable.

$1,382 + 1,744 = 3,126$ milligrams

Check your work.

Since we used addition to find the answer to this problem, we must use subtraction to check our work. If the total weight of the seashells is 3,126, then the total minus the weight of Chris's seashell should equal the weight of Cynthia's seashell: $3,126 - 1,382 = 1,744$. We have chosen the correct operation and answered the question.

7. *Read the entire word problem.*

We are given the total number of cards in a box and the number of cards used by Emma.

Identify the question being asked.

We are looking for the number of cards left in the box.

Underline the keywords.

The word *left* is a keyword.

List the possible operations.

The keyword *left* signals subtraction.

Write number sentences for each operation.

We have two subtraction number sentences we can write using these numbers: 625 − 313 and 313 − 625.

Solve the number sentences and decide which answer is reasonable.

625 − 313 = 312 and 313 − 625 = −312. The number of cards left in the box can't be a negative number, since that would be impossible. Since the box began with 625 cards and 313 cards are used, the number of cards left is the positive difference between those two numbers. The answer, 312 cards, is reasonable.

Check your work.

Since we used subtraction to find the answer to this problem, we must use addition to check our work. If 313 cards are used and 312 cards are left, the total size of the box, 625, should be equal to the sum of 313 and 312: 313 + 312 = 625. We have answered the question correctly.

8. *Read the entire word problem.*

We are given the number of rolls of nickels Ethan has and the number of nickels in each roll.

Identify the question being asked.

We are looking for the number of nickels in 15 rolls.

Underline the keywords.

The word *each* is a keyword.

List the possible operations.

The keyword *each* could signal multiplication or division.

Write number sentences for each operation.

We have one multiplication sentence and two division sentences to write: 12 × 40, 12 ÷ 40, and 40 ÷ 12.

Solve the number sentences and decide which answer is reasonable.

12 × 40 = 480, 12 ÷ 40 = 0.3, and 40 ÷ 12 ≈ 3.3. The decimal answers do not make sense, since the number of nickels in a roll is a whole number and the number of rolls is a whole number. Multiplication makes sense, since we are given the number of nickels in one roll and we are looking for the number of nickels in 12 rolls. The answer, 480, is reasonable.

Check your work.

Since we used multiplication to find the answer to this problem, we must use division to check our work. If there are 480 nickels in 12 rolls, then the number of nickels in one roll is equal to 480 divided by 12: 480 ÷ 12 = 40. We have chosen the correct operation and answered the question.

9. *Read the entire word problem.*

We are given the total number of minutes of footage and the number of days over which the footage was recorded.

Identify the question being asked.

We are looking for the number of minutes each day.

Underline the keywords.

The word *each* is a keyword.

List the possible operations.

The keyword *each* tells us we'll need to use either multiplication or division.

Write number sentences for each operation.

We have one multiplication number sentence: 1,565 × 5. We have two division number sentences: 1,565 ÷ 5 and 1,565 ÷ 5.

Solve the number sentences and decide which answer is reasonable.

1,565 × 5 = 7,825; 1,565 ÷ 5 = 313, 5 ÷ 1,565 ≈ 0.003. David records a total of 1,565 minutes of footage in five days, which means that the number of minutes recorded each day is fewer than 1,565, so multiplication is not the right operation. The number 0.003 is too small to be the number of minutes David recorded each day, because at that rate, he'd have to record for 1,000 days just to record three minutes of footage. The answer, 313 minutes, is a reasonable number of minutes for each of the five days, since it is less than 1,565 but not too small.

Check your work.

Since we used division to find the answer to this problem, we must use multiplication to check our work. If David records 313 minutes each day for five days, then the total number of minutes of footage is equal to the product of 313 and 5: 313 × 5 = 1,565. We have chosen the correct operation and answered the question.

10. *Read the entire word problem.*

We are given the starting weight of the polar bear and the number of pounds the bear loses.

Identify the question being asked.

We are looking for the weight of the bear now.

Underline the keywords.

None of the keywords we learned in Chapter 1 appear in this problem, so we will have to use the context of the word problem.

List the possible operations.

The polar bear loses 45 pounds over the course of the year, which means that the bear's weight decreases. We need to use an operation that lowers the number 903, such as subtraction or division.

Write number sentences for each operation.

We have two subtraction number sentences and two division sentences: $903 - 45$, $45 - 903$, $903 \div 45$, and $45 \div 903$.

Solve the number sentences and decide which answer is reasonable.

$903 - 45 = 858$, $45 - 903 = -858$, $903 \div 45 \approx 20.067$, and $45 \div 903 \approx 0.05$. The weight of the bear can't be negative, so the answer -858 is wrong. The bear would have had to lose almost all of its weight to weigh either 20.067 or 0.05 pounds. The only reasonable answer is 858 pounds. This weight is 45 pounds less than the starting weight of the bear.

Check your work.

Since we used subtraction to find the answer to this problem, we must use addition to check our work. If the bear weighs 858 pounds after losing 45 pounds, then the original weight of the bear is equal to the sum of its original weight and the number of pounds it lost: $858 + 45 = 903$ pounds. We have chosen the correct operation and answered the question.

11. *Read the entire word problem.*

We are given the amount of time Monica rides the subway, the amount of time she walks, and the speed of the subway.

Identify the question being asked.

We are looking for the total time it takes Monica to get to work.

Underline the keywords.

There are no keywords in this problem, so we will have to use the context of the word problem.

Cross out extra information.

Since we are looking for the total time it takes Monica to get to work, we don't need the average speed of the subway. Cross out that part of the word problem.

List the possible operations.

To find the time it takes Monica to get to work, we need to combine the time she spends on the subway and the time she spends walking. We need to use addition.

Write number sentences for each operation.

We have just one addition sentence: 35 + 13.

Solve the number sentences and decide which answer is reasonable.

35 + 13 = 48. Since this number is not too large, but still greater than both the time Monica spends on the subway and the time she spends walking, this answer is reasonable.

Check your work.

Since we used addition to find the answer to this problem, we must use subtraction to check our work. If it takes Monica 48 minutes to get to work, then the number of minutes she spends walking should be equal to the total time minus the number of minutes she spends on the subway: 48 − 35 = 13 minutes. We've chosen the correct operation and answered the question.

12. *Read the entire word problem.*

We are given the cost of a stamp, the number of stamps Lisa buys, and the number of packages she mails.

Identify the question being asked.

We are looking for the amount of money Lisa spends on stamps.

Underline the keywords.

There are no keywords in this problem, so we will have to use the context of the word problem.

Cross out extra information.

Since we are looking for the amount of money Lisa spends on stamps, we don't need to know how many packages she sent. Cross out that part of the word problem.

List the possible operations.

The cost of 25 stamps is greater than the cost of one stamp. We need to choose an operation that makes the number 41 larger, such as addition or multiplication.

Write number sentences for each operation.

We have one addition number sentence and one multiplication sentence: 41 + 25 and 41 × 25.

Solve the number sentences and decide which answer is reasonable.

$41 + 25 = 66$ and $41 \times 25 = 1{,}025$. The answer 66 cents is too small; the cost of two stamps would be more than 66 cents, since $41 + 41 = 82$. Multiplying the number of stamps by the price of one stamp is the correct operation. Lisa spends 1,025 cents on stamps.

Check your work.

Since we used multiplication to find the answer to this problem, we must use division to check our work. If the cost of 25 stamps is 1,025 cents, then the cost of one stamp is equal to 1,025 divided by 25: $1{,}025 \div 25 = 41$. We've chosen the correct operation and answered the question.

13. *Read the entire word problem.*

We are given the distance each girl flies and the time each girl spends flying.

Identify the question being asked.

We are looking for the difference between the distances the girls fly.

Underline the keywords.

The phrase *farther than* (although it is split by other words) signals subtraction.

Cross out extra information.

Since we are looking for the difference between the distances the girls fly, we don't need to know how long they spent flying. Cross out the numbers of hours.

List the possible operations.

To find how much farther Lindsay flies than Jamie flies, we must use subtraction.

Write number sentences for each operation.

We have two subtraction sentences: $2{,}462 - 719$ and $719 - 2{,}462$.

Solve the number sentences and decide which answer is reasonable.

$2{,}462 - 719 = 1{,}743$ and $719 - 2{,}462 = -1{,}743$. The difference between their distances cannot be negative, so 1,743 must be the correct answer. It is less than the larger of the two distances, so it is a reasonable answer.

Check your work.

Since we used subtraction to find the answer to this problem, we must use addition to check our work. If Lindsay flew 1,743 miles more than

Jamie, then the sum of that distance and the distance Jamie flew must be equal to the distance Lindsay flew: 1,743 + 719 = 2,462 miles. We've chosen the correct operation and answered the question.

14. *Read the entire word problem.*

We are given the number of seats in the hall, the number of seats sold, and the cost of one ticket.

Identify the question being asked.

We are looking for the total amount of money collected.

Underline the keywords.

The word *per* is a keyword.

Cross out extra information.

Since we are looking for the amount of money collected, we need only the number of seats sold and the cost of a ticket. We don't need to know how many seats there are in the concert hall, so cross out that number.

List the possible operations.

The keyword *per* can signal multiplication or division.

Write number sentences for each operation.

We have one multiplication sentence and two division sentences: 3,216 × 18; 3,216 ÷ 18; and 18 ÷ 3,216.

Solve the number sentences and decide which answer is reasonable.

3,216 × 18 = 57,888; 3,216 ÷ 18 ≈ 178.67; and 18 ÷ 3,216 ≈ 0.006. The results of the division number sentences are not reasonable. The total amount of money collected will be more than the cost of one ticket because more than one ticket was sold. Since we are given the cost of one ticket and need to find the cost of many tickets, we must use multiplication. The answer, $57,888, is reasonable.

Check your work.

Since we used multiplication to find the answer to this problem, we must use division to check our work. If the total collected by the concert hall is $57,888, then that total divided by the ticket price, $18, must equal the number of seats sold: $57,888 ÷ $18 = 3,216. We've chosen the correct operation and answered the question.

15. *Read the entire word problem.*

We are given the number of students who attended last year, the increase in enrollment, and the cost of tuition.

Identify the question being asked.

We are looking for the number of students attending this year.

Underline the keywords.

The word *increase* is a keyword.

Cross out extra information.

Since we are looking for the number of students attending this year, we don't need to know the increase in the cost of tuition. Cross out that number.

List the possible operations.

The keyword *increase* signals addition. We need to add the number of students who attended last year to determine the size of the increase.

Write number sentences for each operation.

We have just one addition sentence: 2,314 + 239.

Solve the number sentences and decide which answer is reasonable.

2,314 + 239 = 2,553. The new number of students attending is larger than the old number of students who attended the college, so this answer is reasonable.

Check your work.

Since we used addition to find the answer to this problem, we must use subtraction to check our work. If the new number of students attending Lincoln College is 2,553, then that number minus the increase should be equal to the number of students who attended the college last year: 2,553 − 239 = 2,314. We've chosen the correct operation and answered the question.

16. *Read the entire word problem.*

We are given a number of feet.

Identify the question being asked.

We're looking for the number of inches in those feet.

Underline the keywords.

There are no keywords in this problem, so we will have to use the context of the problem.

Cross out extra information and translate words into numbers.

There is no extra information in this word problem. There is only one number, 9, but there are 12 inches in a foot. The words *are in* mean that we must convert feet into inches.

List the possible operations.

To convert feet, a larger unit of measure, to inches, a smaller unit of measure, we must multiply.

Write number sentences for each operation.

We have just one multiplication sentence: 12 × 9.

Solve the number sentences and decide which answer is reasonable.

12 × 9 = 108 inches. By using the number of inches in one foot to find the number of inches in 9 feet, we would expect our answer to be much larger than 9. The answer, 108 inches, is reasonable.

Check your work.

Since we used multiplication to find the answer to this problem, we must use division to check our work. Since there are 108 inches in 9 feet, dividing the number of inches by the number of feet should give us the number of inches in one foot, 12: 108 ÷ 9 = 12. We've chosen the correct operation and answered the question.

17. *Read the entire word problem.*

We are given the weight and cost of a dozen books.

Identify the question being asked.

We're looking for the cost of one book.

Underline the keywords.

The word *per* is a keyword.

Cross out extra information and translate words into numbers.

We are looking for the cost per book, so we do not need the weight of the books. Cross out that number. That leaves us with just the number $360. The word *dozen* is equal to 12. Replace the word *dozen* with 12 in the problem.

List the possible operations.

The keyword *per* can signal multiplication or division.

Write number sentences for each operation.

We have one multiplication sentence and two division sentences: $360 × 12, 360 ÷ 12, and 12 ÷ $360.

Solve the number sentences and decide which answer is reasonable.

$360 × 12 = $4,320; $360 ÷ 12 = $30; and 12 ÷ $360 ≈ $0.03. Since the cost of 12 books is $360, the cost of one book must be less than that, so multiplication is the wrong operation. A cost of $0.03 is not reasonable, since it would take thousands of books to total $360. The answer, $30, is reasonable. Dividing the total cost by the number of books gives us the cost per book.

Check your work.

Since we used division to find the answer to this problem, we must use multiplication to check our work. If the cost per book is $30, and there are 12 books, the product of $30 and 12 should equal the total cost of the books, $360: 12 × $30 = $360. We've chosen the correct operation and answered the question.

18. *Read the entire word problem.*

We are given the number of pairs of socks Nate has.

Identify the question being asked.

We are looking for the number of individual socks.

Underline the keywords.

There are no keywords in this word problem.

Cross out extra information and translate words into numbers.

There is no extra information in this problem, but the word *pair* represents the number 2. A pair of socks is two individual socks.

List the possible operations.

We have the number of pairs of socks, and there are two socks in a pair, so we must use multiplication to find the number of individual socks.

Write number sentences for each operation.

We have one multiplication sentence: 18 × 2.

Solve the number sentences and decide which answer is reasonable.

18 × 2 = 36. Since there are two socks in a pair, it makes sense that the number of individual socks is twice the number of pairs.

Check your work.

Since we used multiplication to find the answer to this problem, we must use division to check our work. The number of individual socks, 36, divided by the number of socks in a pair, 2, should equal the number of pairs of socks, 18: 36 ÷ 2 = 18 pairs of socks. We've chosen the correct operation and answered the question.

19. *Read the entire word problem.*

We are given number of sides of a cube that are painted red.

Identify the question being asked.

We are looking for the number of sides that are not painted red.

Underline the keywords.

There are no keywords in this word problem.

Cross out extra information and translate words into numbers.

There is no extra information in this problem, but a cube has six sides. We need to use the total number of sides of a cube to help us find the number of sides that are not painted red.

List the possible operations.

We have the number of sides of a cube and the number of sides that are painted red. By subtracting the number of red sides from the total number of sides, we can find the number of sides that are not red.

Write number sentences for each operation.

We know that we need to subtract the number of red sides from the total number of sides, so we have only one number sentence: $6 - 4$.

Solve the number sentences and decide which answer is reasonable.

$6 - 4 = 2$ sides. A cube has six sides, and if four are painted red, then the other two are not.

Check your work.

Since we used subtraction to find the answer to this problem, we must use addition to check our work. The number of red sides, 4, plus the number of sides that are not red, 2, should equal the total number of sides, 6: $4 + 2 = 6$ sides. We've chosen the correct operation and answered the question.

20. *Read the entire word problem.*

We are given the number of cups of milk and orange juice that Dylan drinks.

Identify the question being asked.

We are looking for the number of pints of milk Dylan drinks.

Underline the keywords.

There are no keywords in this word problem.

Cross out extra information and translate words into numbers.

We are looking for the number of pints of milk that Dylan drinks, so cross out the number of cups of orange juice he drinks. That leaves just the number of cups of milk. There are 2 cups in a pint, and we can use that number to find out how many pints of milk Dylan drank.

List the possible operations.

A pint is larger than a cup, so to convert cups to pints, we must divide the number of cups by the number of cups in a pint, 2.

Write number sentences for each operation.

We know that we need to divide the number of cups of milk Dylan drank by 2, so we have only one number sentence: $4 \div 2$.

Solve the number sentences and decide which answer is reasonable.

$4 \div 2 = 2$. Since a pint is a larger unit of measure than a cup, we would expect the number of pints to be smaller than the number of cups. The answer, 2, is reasonable.

Check your work.

Since we used division to find the answer to this problem, we must use multiplication to check our work. The number of pints Dylan drank, 2, multiplied by the number of cups in a pint, 2, should be equal to the number of cups Dylan drank, 4: $2 \times 2 = 4$ cups. We've used the correct operation and answered the question.

Using Pictures, Tables, and Venn Diagrams

We can apply the eight-step process we developed in the last chapter to solve any word problem, but sometimes it is easier to draw a picture or create a table. Pictures and tables can help us see exactly what a question is asking us.

DRAWING A PICTURE

Sketching a word problem can help us decide which operation to use and what numbers to plug into an equation.

Example

A piano has 88 keys that are either black or white. If 36 keys are black, how many are white?

First, let's use the eight-step process:

1. *Read the entire word problem.*
   We are given the total number of keys on a piano and the number of keys that are black.
2. *Identify the question being asked.*
   We're looking for the number of keys that are white.
3. *Underline the keywords.*
   There are no keywords in this problem.

4. *Cross out extra information and translate words into numbers.*

There is no extra information to cross out, and there are no words that need to be translated into numbers.

So far, the eight-step process hasn't moved us much closer to solving the problem. Let's try drawing a picture. It's fine if you're not an artist; the picture just needs to represent accurately the problem. We are told that the piano has 88 keys, so start by drawing a piano with 88 keys:

We know that all the keys are either black or white, and we know that 36 keys are black, so color 36 of the keys black. It doesn't matter if the picture looks exactly the way a real piano would look; we just need 36 of the 88 keys colored black:

Now we can see that the remaining keys on the right side of the piano are the white keys. By counting them, we can find that there are 52 white keys.

After coloring 36 of the keys black, you may have realized that this was a subtraction problem. The number of white keys is equal to the total number of keys minus the number of black keys: 88 − 36 = 52. The picture is meant to help you as far as you need help. Once you realize what must be done to solve a problem, you can leave your picture and find the answer.

Example

A bookshelf has six shelves. If there are 16 books on each shelf, how many books are on the bookshelf?

The keyword *each* tells us that this is likely either a multiplication or division problem, but a picture may help us solve this problem faster than the eight-step process. Draw a bookshelf with six shelves and put 16 books on each shelf:

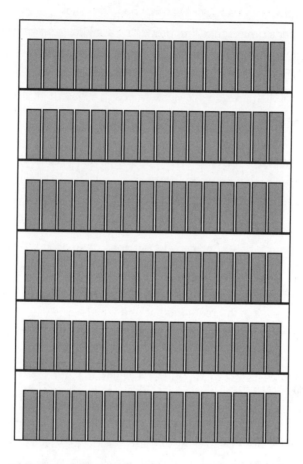

We can now see that there are many books on this shelf—many more than 16. We won't use division or subtraction; we need an operation that increases 16 by 6 times. Counting the books on the shelf would give us the total number of books, but the bookshelf itself represents 16 multiplied by 6: 16 × 6 = 96. There are 96 books on the shelf.

PRACTICE LAP

DIRECTIONS: Solve these word problems by drawing a picture.

1. Mike has 14 marbles in a sack. If he puts nine more marbles in the sack, how many marbles will be in the sack?

2. Jennifer's dresser has four drawers. Each drawer contains eight shirts. How many shirts are in Jennifer's dresser?

3. There are five classrooms on the third floor of Nelson Middle School. If there are 22 desks in each classroom, how many desks are on the third floor of the school?

4. A jar holds 35 candies. If 19 of the candies are lemon and the rest are strawberry, how many strawberry candies are in the jar?

5. Jasmine baked 48 cookies. If she gives three cookies to each of her friends, to how many friends did she give cookies?

6. A tube can hold four tennis balls. If Kai needs to put 36 tennis balls into tubes, how many tubes does he need?

7. There are 15 tables in a restaurant and three waitresses. Six people can sit at each table. What is the greatest number of people who can sit in the restaurant at one time?

8. A pitcher holds 40 ounces of water. How many ounces would six pitchers hold?

9. Melissa's quilt is made up of 54 squares. If she sews 18 more squares onto the quilt, how many squares will be on the quilt?

10. There are 63 seats in the cafeteria. If 38 people are seated in the cafeteria, how many seats are free?

BUILDING A TABLE

We don't always need to draw a picture. Often, a table can be just as helpful. In the bookshelf example, we drew six shelves and placed 16 books on each shelf. We could have drawn a table and placed the numbers of books in the table:

Shelf Number	Number of Books
1	16
2	16
3	16
4	16
5	16
6	16

The table, just like the picture, helps us to see that we have six sets of 16. The best way to find the total number of books is to multiply 6 by 16: $6 \times 16 = 96$ books.

The type of table we build depends on the information in the word problem. Tables are especially good for organizing multiplication problems, as we just saw. But they are also good for solving other kinds of problems.

Example

A carnival game has red pins and blue pins. For every red pin that is knocked down, a player scores three points, and for every blue pin that is knocked down, a player scores five points. Fred knocked down eight pins and scored 34 points. How many of each pin did Fred knock down?

We can tell right away that this isn't a typical addition, subtraction, multiplication, or division problem. We know that Fred knocked down eight pins, so the number of red pins plus the number of blue pins must equal eight. We also know that Fred scored 34 points, so the number of points scored from red pins plus the number of points scored from blue pins must equal 34.

Let's say Fred knocked down only red pins. Since each red pin is worth three points, Fred would have scored $8 \times 3 = 24$ points. But Fred scored 34 points, so he must have knocked down some blue pins. By putting the information in a table, we can make different combinations of red and blue pins until we find the right total. The number of red pins times three plus the number of blue pins times five must equal 34. The following table shows all the possible combinations of eight pins that Fred could have knocked down:

Number of Red Pins	Number of Blue Pins	Number of Points from Red Pins (number of red pins times 3)	Number of Points from Blue Pins (number of blue pins times 5)	Total
8	0	24	0	24
7	1	21	5	26
6	2	18	10	28
5	3	15	15	30
4	4	12	20	32
3	5	9	25	34
2	6	6	30	36
1	7	3	35	28
0	8	0	40	40

The sixth row of the table shows Fred scoring a total of 34 points. To score 34 points, he had to knock down three red pins and five blue pins. Solving this problem without a table would have been tough!

INSIDE TRACK

WHEN BUILDING A table, you can stop filling in numbers once you have found the answer. In the last example, we could have stopped after completing row 6. Remember, the table is a tool for helping you find the answer; completing the table is not the goal.

Example

Ice cream cones cost $2, and ice cream cups cost $3. Dom sells a total of ten cones and cups and collects $23. How many cones and cups did Dom sell?

We can build a table similar to the one we built in the last example. The amount of money Dom collected is equal to the number of cones sold multiplied by $2 plus the number of cups sold multiplied by $3. If Dom sold only cones, he would have collected 10 × $2 = $20, and if he had sold only cups, he would have collected 10 × $3 = $30. Since Dom collected $23, he must have sold some of each:

Number of Cones	Number of Cups	Money Collected from Selling Cones	Money Collected from Selling Cups	Total
10	0	$20	0	$20
9	1	$18	$3	$21
8	2	$16	$6	$22
7	3	$14	$9	$23

We can stop building our table—we've found our answer. In order to collect $23 from ten cones and cups, Dom must have sold seven ice cream cones and three ice cream cups.

Some word problems don't involve numbers at all. Building a table can help us solve these types of problems, too.

Example

Nicholas is older than Anthony but younger than Louis. Marie is older than Louis but younger than Jill. Who is the oldest, and who is the youngest?

Build a table ordered from oldest to youngest from top to bottom. Start by placing Nicholas in the middle of the table with Anthony beneath him and Louis above him:

Louis
Nicholas
Anthony

Next, add Marie to the table above Louis, and place Jill above Marie, since Jill is older than Marie:

Jill
Marie
Louis
Nicholas
Anthony

It's easy to see now that Jill is the oldest and Anthony is the youngest. The table we used to solve this problem was simple, but it helped us organize the information given in the word problem.

PRACTICE LAP

11. Rico has seven coins in his pocket, all of which are nickels or quarters. If Rico has $0.95 in his pocket, how many nickels and how many quarters are in his pocket?

12. Tickets to a movie cost $4 for children and $7 for adults. If 15 people see the movie and $87 is paid, how many children and how many adults saw the movie?

13. Five people are standing in line. Katie is ahead of Danny, who is ahead of Kenny but behind Casey. Coral is behind Kenny. If Casey is not first in line, who is?

14. There are six teams in the Eastern Volleyball League. The Saints have 16 wins, the Cardinals have 20 wins, the Comets have eight wins, the Stars have 12 wins, the Bees have 11 wins, and the Tigers have 18 wins. How many games must the Bees win to tie for third place?

15. Harold is taller than Mickey but shorter than Peter. Peter is taller than Joanna but shorter than Dennis. If Joanna is taller than Mickey, which two people could be the same height?

PACE YOURSELF

MAKE A TABLE with the heights and ages of all your friends. Which one of your friends is the oldest? The tallest? What is the difference in age between your oldest friend and your youngest friend? Your tallest friend and your shortest friend? Use your table to answer these questions.

VENN DIAGRAMS

Pictures and tables can help us solve problems, and one specific type of drawing, a Venn diagram, can help us solve word problems, too.

A **Venn diagram** is a drawing of one or more shapes that each represents a quantity of an item. The shapes can overlap to show items that are members of more than one set.

The following Venn diagram shows six people who own either a cat, a dog, or both:

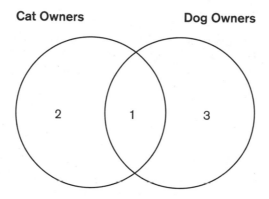

By looking at this diagram, we can see that two people own only a cat, because the number 2 is in the circle that represents Cat Owners. Three people own only a dog, because there is a 3 in the circle labeled Dog Owners. The overlapping area of the two circles represents people who own both a cat and a dog. Since there is a 1 in this area, that means that one person owns both a cat and a dog.

CAUTION

READING A VENN diagram is not hard, but creating one can be a little tricky. Because areas of Venn diagrams overlap, we must be careful not to count the same items more than once or not at all. As the last example, a word problem may tell us that four people own at least a dog, three people own at least a cat, and one person owns both. Reading quickly, you may think that this problem is talking about eight people (4 + 3 + 1). However, one person—the person who owns both a cat and a dog—has been described three times in that sentence. That person is someone who owns at least a dog, someone who owns at least a cat, and someone who owns both. Creating a Venn diagram helps us recognize that an item can belong to many sets and prevents us from counting that item more than once.

Example

The 71 seventh-grade students at Lakeland Middle School can take biology, chemistry, or both, but they must take at least one of the two. If there are 46 students taking biology and 39 students taking chemistry, how many are taking only biology?

We'll need to draw two overlapping circles: one to represent the students taking biology and one to represent the students taking chemistry. The area where the circles overlap will represent the number of students taking both biology and chemistry:

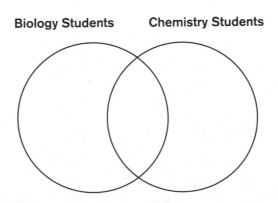

Biology Students   **Chemistry Students**

We can't place "46" in the biology circle, because we must separate that 46 into two numbers: the number of students who take just biology and the number of students who take both biology and chemistry. In the same way, we must separate the 39 chemistry students into the number of students taking only chemistry and the number of students taking both biology and chemistry.

Add the number of biology and chemistry students together: 46 + 39 = 85. However, there are only 71 students in the seventh grade: 85 − 71 = 14. When we added 46 and 39, we must have counted 14 students twice. Therefore, 14 students are taking both biology and chemistry:

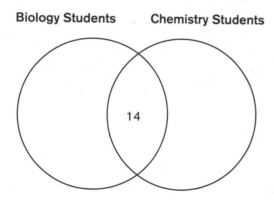

Since there are 46 students taking biology, and 14 taking both biology and chemistry, then there are 46 − 14 = 32 students taking only biology (and 39 − 14 = 25 students taking only chemistry):

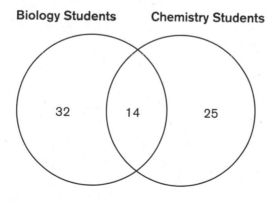

Venn diagrams can have more than two regions—they can have any number of regions, with any number of them overlapping. We can use systems of equations to solve these kinds of questions.

Example

The most popular sports at Lakeland Middle School are soccer, lacrosse, and basketball. There are 20 students on the soccer team, 31 students on the lacrosse team, and 16 students on the basketball team. Also, 12 students play only soccer, 19 students play only lacrosse, and six students play only basketball. If 50 students play sports, how many play all three sports? How many play just soccer and lacrosse? How many play just soccer and basketball? How many play just lacrosse and basketball?

We start by drawing three overlapping circles. We must draw them as shown here, so that there is an area where just soccer and lacrosse overlap, an area where just lacrosse and basketball overlap, an area where just soccer and basketball overlap, and an area where all three overlap. We are given the number of students who play only one sport for each sport, so we can label those areas of the diagram. The unknown areas, the areas that represent students who play multiple sports, are labeled with variables:

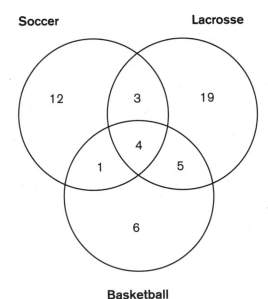

We know that 50 students play sports and that $12 + 6 + 19 = 37$ students play only one sport. That means that $50 - 37 = 13$ students play more than one sport. The overlapping areas, labeled a, b, c, and d respectively, represent these 13 students: $a + b + c + d = 13$. We also know that there are 20 students on the soccer team, which means that $a + b + d + 12 = 20$. In the same way, there are 31 students on the lacrosse team, so $a + c + d + 19 = 31$, and the number of soccer players can be represented with $b + c + d + 6 = 16$.

How can we use these equations? By combining different pairs of equations, we can find the value of each variable: $a + b + d + 12 = 20$, which means that $a + b + d = 8$, if we subtract 12 from both sides of the equation. Now compare this equation to the first equation:

$$a + b + c + d = 13$$
$$a + b + d = 8$$

If we subtract the second equation from the first, we are left with $c = 5$. Checking our Venn diagram, we see that c represents the number of students who play only lacrosse and basketball. To find a, the number of students who play only soccer and lacrosse, let's combine two equations again. Since $b + c + d + 6 = 16$, $b + c + d = 10$ after we subtract 6 from both sides of the equation.

$$a + b + c + d = 13$$
$$b + c + d = 10$$

Again, subtract the second equation from the first. We are left with $a = 3$. Three students play only soccer and lacrosse. Now that we have a and c, we can find d using the equation $a + c + d + 19 = 31$. Substitute 3 for a and 5 for c: $3 + 5 + d + 19 = 31$, and $d = 4$. There are four students who play all three sports. That just leaves b, the number of students who play only soccer and basketball. Since $a + b + c + d = 13$, substitute the values of a, c, and d and solve for b: $3 + b + 5 + 4 = 13$, so $b = 1$. There is one student who plays soccer and basketball.

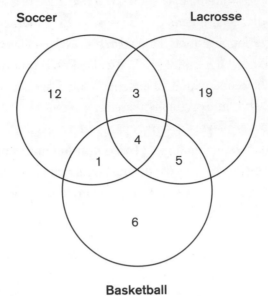

Soccer   Lacrosse

Basketball

16. Seventy-eight hot dogs and 60 hamburgers are sold at a carnival booth. If 121 people ate (or bought) one hot dog, one hamburger, or one of each, how many people bought one of each?

17. A vendor sells hats and shirts. 24 people buy both a hat and a shirt. If the vendor sells a total of 76 hats and 111 shirts, and no one buys more than one of each, how many customers does the vendor have?

18. The zoo has exhibits in the reptile room, the bird sanctuary, and the aquarium. Nineteen people visited only the reptile room, and 37 people visited only the aquarium. Twenty people saw two exhibits, but no one saw all three. If 54 people visited the aquarium, how many people went to the reptile room *and* the bird sanctuary?

19. A survey asked 40 people about the types of movies they like. Eleven people like action movies, but not comedies or horror movies; five people like comedies and horror movies, but not action movies; and two people like all three types of movies. If 17 people like horror movies, what is the greatest number of people who like only horror movies?

PATTERNS

While tables, pictures, and Venn diagrams can help us solve word problems, sometimes all we need is a simple list. When a word problem involves a pattern, sequence, or series, making a list can help us find the answer.

Example

> The volume of water in a tank doubles every hour. If there are 16 gallons in the tank now, how many gallons will be in the tank four hours from now?

We could solve this problem using algebra, but making a list may be easier. Start by writing the number of gallons in the tank now:

16

The volume of water doubles every hour. The word *doubles* means "multiply by 2." After one hour, there will be $16 \times 2 = 32$ gallons in the tank. To find the number of gallons after two hours, multiply 32 by 2. Continue to double each number:

16, 32, 64, 128, 256

After four hours, there will be 256 gallons of water in the tank.

Example

> Erle has $140 in his bank account. Every week he adds $20 to the account and then spends one-fourth of the total. How much money does Erle have in his account after three weeks?

Erle starts with $140. Add $20 and divide by 4: $140 + $20 = $160, $160 ÷ 4 = 40. However, $40 is not what Erle has left; that is how much he spends. If Erle spends $40, he will have $160 − $40 = $120. The pattern is the same for the second week: $120 + $20 = $140, $140 ÷ 4 = $35, $140 − $35 = $105. In week three, Erle adds $20 to $105: $20 + $105 = $125. Check your figures: $20 + $105 = $125; 125 ÷ 4 = $31.25; $125 − 31.25 = $93.75. Erle's totals for three weeks are:

$140, $120, $105, $93.75

20. William buys two books every time he visits the bookstore. He visits the bookstore every other day, and he went to the bookstore today. How many books will he have after 15 days?

21. Cassie can type 65 words per minute. Every three minutes, she deletes 20 words from her document. How many words are in her document after 12 minutes?

22. After a new exhibit opens at the science museum, the number of visitors increases by four times each day for three consecutive days. If 30 people came to the museum the day before the exhibit opened, how many people came to the museum three days later?

23. A tennis tournament plans to give out 1,024 tennis balls by placing them in a large container and allowing fans to take from the container. If the number of balls decreases by four times every hour, how many hours will it take before there is only one ball left in the container?

24. Brian has 20 points. Every time he wins a game, his point total triples. Every time he loses, his point total is cut in half. If Brian wins two games, and then loses two games, how many points will he have?

SUMMARY

DRAWING PICTURES OR diagrams and creating tables and lists can make solving word problems easier. By organizing information in these different ways, we're able to get a clear idea of what a problem is asking us and how to find the answer to these different types of problems.

ANSWERS

1. Start by drawing a sack with 14 marbles in it:

Now, draw nine more marbles in the sack:

We can count the marbles in the sack, or you may have realized that this is an addition problem, since nine marbles were added to the sack: 14 + 9 = 23 marbles.

2. Draw a dresser with four drawers, and put eight shirts in each drawer:

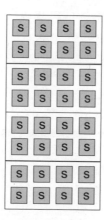

We can count the number of shirts in the picture to find the total. Since there are four drawers, each with eight shirts, you can also find the total number of shirts by multiplying the number of drawers by the number of shirts in each drawer: $4 \times 8 = 32$ shirts.

3. Start by drawing five large classrooms:

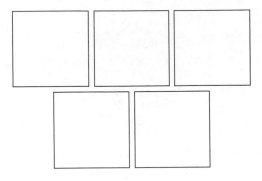

Now, draw 22 desks in each room:

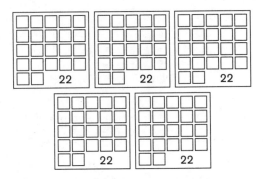

This is a multiplication problem. Since each classroom has 22 desks, and there are five rooms, we can multiply the number of rooms by the number of desks in each room to find the total number of rooms: $5 \times 22 = 110$ desks. We can check our answer by counting the number of desks in our picture.

4. Draw a jar with 35 candies:

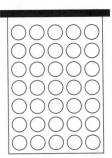

Put an "L" on 19 of the candies so we know that those are lemon candies:

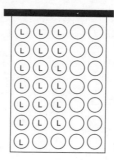

We can either count the unlabeled candies, or we can subtract 19 from 35. The number of candies that are strawberry is equal to the total number of candies minus the number of lemon candies: 35 − 19 = 16 strawberry candies.

5. Draw 48 cookies:

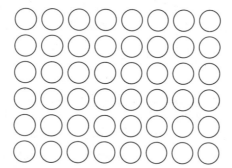

Since Jasmine gave three cookies to each of her friends, draw circles around groups of three cookies:

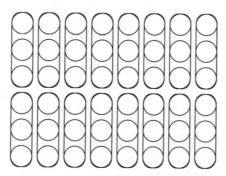

You just drew 16 circles, which means that Jasmine gave cookies to 16 friends. This is a division problem: 48 ÷ 3 = 16 cookies.

6. Draw 36 tennis balls:

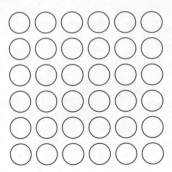

Each tube holds four tennis balls. Draw a circle around groups of four tennis balls:

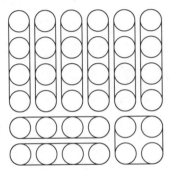

The 36 tennis balls were divided into groups of 4: $36 \div 4 = 9$ tubes. Since we just drew nine circles around groups of tennis balls, we know that our division answer is correct.

7. Draw 15 tables. We are looking for the number of people who can sit in the restaurant, so the number of waitresses is extra information. Cross out that number.

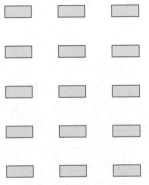

Now, draw six X's at each table to represent the people:

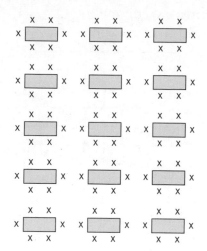

We are given the number of tables and the number of people who can sit at each table, and we are looking for the total number of people who can fit in the restaurant. Count the number of people in our picture, or multiply the number of tables by the number of people who can sit at each table: 15 × 6 = 90 people.

8. Draw six pitchers of water and label each with the number 40:

Add the six 40s in your picture, or multiply 40 by 6: 40 × 6 = 240 ounces.

9. Draw a quilt that has 54 squares:

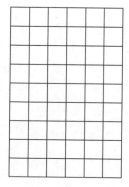

Add 18 more squares onto the quilt:

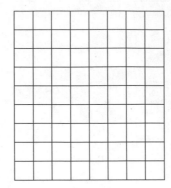

Count the number of squares in the picture, or since you just added 18 squares to the picture, add 18 to 54: 54 + 18 = 72 squares.

10. Draw 63 seats:

Cross out 38 seats, to show that those seats are taken:

Count the number of seats that are not crossed out. We have just subtracted 38 from 63: 63 − 38 = 25 seats.

11. Build a table with columns for the number of nickels, the number of quarters, the value of the nickels, the value of the quarters, and the total value. The total value is equal to the number of nickels multiplied by $0.05 plus the number of quarters multiplied by $0.25, and the total number of coins should always equal 7, since Rico has seven coins in his pocket:

Number of Nickels	Number of Quarters	Value of Nickels	Value of Quarters	Total
7	0	$0.35	$0.00	$0.35
6	1	$0.30	$0.25	$0.55
5	2	$0.25	$0.50	$0.75
4	3	$0.20	$0.75	$0.95

We can stop building our table—we've found our answer. If Rico has seven coins in his pocket, all of which are nickels or quarters and total $0.95, then he has four nickels and three quarters.

12. Build a table with columns for the number of children's tickets, the number of adult tickets, the total collected from children, the total collected from adults, and the overall total. The overall total is equal to the number of children's tickets multiplied by $4 plus the number of adult tickets multiplied by $7, and the total number of tickets should always equal 15, since 15 people saw the movie:

Number of Children's Tickets	Number of Adult Tickets	Total Collected from Children	Total Collected from Adults	Total
15	0	$60	$0	$60
14	1	$56	$7	$63
13	2	$52	$14	$66
12	3	$48	$21	$69
11	4	$44	$28	$72
10	5	$40	$35	$75
9	6	$36	$42	$78
8	7	$32	$49	$81
7	8	$28	$56	$84
6	9	$24	$63	$87

If 15 people see the movie and $87 is collected, then six children's tickets and nine adult tickets were sold.

13. Enter each person into a table. Since Katie is ahead of Danny, place her above Danny:

Katie
Danny

Danny is ahead of Kenny but behind Casey. Place Kenny beneath Danny. Place Casey in the same row as Katie, since we don't know yet if Katie comes before Casey or if Casey comes before Katie:

Katie, Casey
Danny
Kenny

Coral is behind Kenny, so place her at the bottom of the table:

Katie, Casey
Danny
Kenny
Coral

Since we are told that Casey is not first in line, she must be behind Katie, and Katie must be first in line.

14. Place each team in a table sorted by the number of wins:

Team	Number of Wins
Cardinals	20
Tigers	18
Saints	16
Stars	12
Bees	11
Comets	8

In order for the Bees to tie for third place, they must equal the number of wins of the third place team, the Saints. Since the Bees have 11 wins now, they must win 16 – 11 = 5 more games in order to tie for third place.

15. Enter each person into a table. Since Harold is taller than Mickey but shorter than Peter, place Peter above Harold and place Mickey beneath Harold:

Peter
Harold
Mickey

Next, we are told that Peter is taller than Joanna, but shorter than Dennis. Since Peter is at the top of the table, we know that Dennis can be put at the very top now, since he is taller than Peter. Joanna is shorter than Peter, so we know that she goes beneath him, but we do not know if she is shorter or taller than Harold or Mickey, so we put her on the same line as Harold and on the same line as Mickey:

Dennis
Peter
Harold, Joanna
Mickey, Joanna

Finally, we are told that Joanna is taller than Mickey, so we can remove her from the bottom row:

Dennis
Peter
Harold, Joanna
Mickey

Looking at the table, we can see that Harold and Joanna could be the same height.

16. Draw a Venn diagram, with one circle representing hot dogs sold and one circle representing hamburgers sold. Label the area that represents people who bought just a hot dog "*a*," label the area that represents people who bought just a hamburger "*b*," and label the overlapping area, the area that represents people who bought one of each, "*c*."

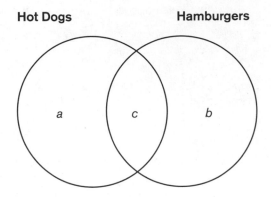

The number of hot dogs sold, $a + c$, is 78, and the number of hamburgers sold, $b + c$, is 60. The total number of hot dogs and hamburgers sold, $a + b + c$, is equal to 121. We can add the first two equations and subtract the third from that total:

$a + c = 78$
$b + c = 60$
$60 + 78 = 138$

$a + b + c = 121$
$138 - 121 = 17$
$b = 17$

Seventeen people bought one of each. Rather than using equations, you may have realized that the number of people who bought both a hamburger and a hot dog was equal to the total number of hot dogs and hamburgers sold minus the number of people: $78 + 60 - 121 = 17$.

17. Draw a Venn diagram, with one circle representing hats sold and one circle representing shirts sold. Label the area that represents people who bought just a hat "*a*," label the area that represents people who bought just a shirt "*b*," and label the overlapping area, the area that represents people who bought one of each, "24."

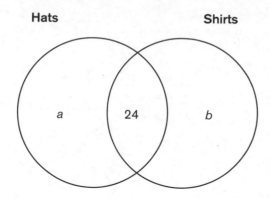

Hats   Shirts

a   24   b

The total number of hats sold is 76, so $a + 24 = 76$; $a = 52$. The total number of shirts sold is 111, so $24 + b = 111$; $b = 87$. The total number of customers is equal to the sum of the values in the Venn diagram: $52 + 24 + 87 = 163$. You may also have realized that the number of customers was equal to the total number of sales minus the number of people who bought both, since those people would have been counted twice in the sum of the number of hats and shirts: $76 + 111 - 24 = 163$.

18. Draw a Venn diagram, with one circle representing visitors to the reptile room, one circle representing visitors to the bird sanctuary, and one circle representing visitors to the aquarium. Label the area that represents people who visited the reptile room and the bird sanctuary "a," label the area that represents people who visited the reptile room and the aquarium "b," label the area that represents the people who visited the bird sanctuary and the aquarium "c," and label the area where all three circles overlap "0," since no one visited all three exhibits. We are also given the number of people who visited only the reptile room (19) and the number of people who visited only the aquarium (37), so place those numbers in the diagram as well:

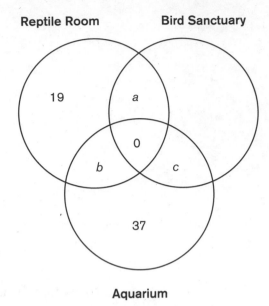

Reptile Room   Bird Sanctuary

Aquarium

Since 54 people visited the aquarium, $b + c + 37 = 54$, and $b + c = 17$. Twenty people saw two exhibits, which means that $a + b + c = 20$. We can subtract the first equation from the second to find a, the number of people who went to the reptile room and the bird sanctuary:

$$a + b + c = 20$$
$$\underline{b + c = 17}$$
$$a = 3$$

Three people went to the reptile room and the bird sanctuary.

19. Draw a Venn diagram, with one circle representing fans of action movies, one circle representing fans of comedies, and one circle representing fans of horror movies. Label the area that represents people who like action movies "11," label the area that represents people who like comedies and horror movies "5," and label the area where all three circles overlap "2," since two people like all three types of movies:

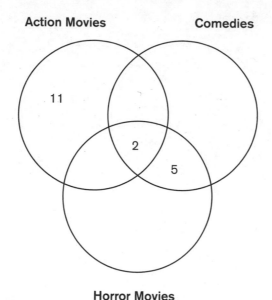

Action Movies   **Comedies**

11

2

5

Horror Movies

Since 17 people like horror movies, the sum of the numbers within the horror movies circle must equal 17. Since two people like all movies and five people like comedies and horror movies, there are seven people who like horror movies and other movies. Therefore, there are $17 - 7 = 10$ people who fall into one of two categories. The first category is people who like only action and horror movies. The second category is people who like only horror movies. The number of people who like only horror movies will be the greatest if there is no one who likes only action and horror movies. Then, all ten would be people who like only horror movies. The Venn diagram that represents this case is shown here:

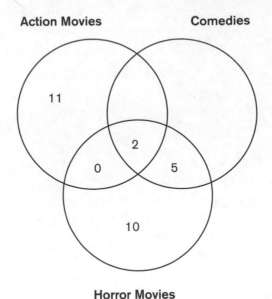

Action Movies   Comedies

Horror Movies

20. Make a list of the number of books William buys each day. Since he went to the store today, he bought two books today. He goes to the bookstore only every other day, which means he will buy zero books tomorrow. Alternate writing 2 and 0 until you have written 15 numbers:

2, 0, 2, 0, 2, 0, 2, 0, 2, 0, 2, 0, 2, 0, 2

William buys two books on each of eight days: $2 \times 8 = 16$ books.

21. Cassie types 65 words each minute for 12 minutes, so list 65 twelve times:

65, 65, 65, 65, 65, 65, 65, 65, 65, 65, 65, 65

Cassie deletes 20 words every third minute, so place −20 under every third number:

65, 65, 65, 65, 65, 65, 65, 65, 65, 65, 65, 65
     −20,   −20,   −20,   −20

Total the 65s and the −20s: $65 \times 12 = 780$, $(-20) \times 4 = -80$. $780 - 80$ $= 700$. Cassie's document has 700 words in it after 12 minutes.

22. The number of visitors on the day before the exhibit opened is 30. Multiply that number by 4, and then multiply each result by 4:

Before the exhibit opened: 30

First day: $30 \times 4 = 120$

Second day: $120 \times 4 = 480$

Third day: $480 \times 4 = 1{,}920$

Three days after the exhibit opened, 1,920 people came to the museum.

23. The container begins with 1,024 tennis balls in it. We divide that number by 4 until we are left with only one ball:

$\frac{1{,}024}{4} = 256$, $\frac{256}{4} = 64$, $\frac{64}{4} = 16$, $\frac{16}{4} = 4$, $\frac{4}{4} = 1$

Since we divided by 4 five times, it took five hours for there to be only one ball left in the container.

24. Brian begins with 20 points, and then he wins two games. Each time he wins, his point total triples: $20 \times 3 = 60$, $60 \times 3 = 180$. Then, Brian loses two games. Each time he loses, his point total is cut in half: $\frac{180}{2} = 90$, $\frac{90}{2} = 45$ points. Brian's total after each game is: 60, 180, 90, 45. After four games, Brian has 45 points.

Word Problem Pitfalls

We've seen how to translate keywords into operations and how to decide which operation to use based on the context of a word problem. However, sometimes keywords and phrases can be misleading. In this chapter, we'll look at how word problems try to trick you—and how you can avoid being tricked.

BACKWARD PHRASES

A **backward phrase** is a group of words and numbers that describes an operation in which the numbers are given in the opposite order that they will appear in a number sentence. For instance, *3 less than 4* is a backward phrase because when the numbers 3 and 4 are placed in a number sentence, 4 appears before 3: 4 – 3.

The phrase *add 4 and 6* can be converted into the number sentence 4 + 6. The numbers fit right into the number sentence as they appear in the phrase. What about the phrase *subtract 4 from 6*? At first glance, you may want to write 4 – 6. Read the phrase again carefully: subtract 4 *from* 6. We are starting with 6 and taking away 4. This phrase, written as a number sentence, is 6 – 4. This kind of phrase is a backward phrase.

Example

If nine pencils are added to a box that holds eight pencils, how many pencils are in the box?

The number 9 is added to 8, the original quantity. The phrase *added to* is a backward phrase. There were already eight pencils in the box, so we are adding 8 and 9: $8 + 9 = 17$ pencils. As we've already learned, addition is commutative, so even if we had added 8 to 9, we still would have found the same answer: $9 + 8 = 17$. However, not all operations are commutative.

Example

North has 14 fewer photos than Jackie, who has 63 photos. How many photos does North have?

The words *fewer* and *than* may be separated, but these words signal that the order of the numbers given in the word problem will need to be switched. *Fewer* is a keyword that signals subtraction, so to find how many photos North has, begin with 63 and subtract 14: $63 - 14 = 49$ photos.

INSIDE TRACK

SUBTRACTION WORD PROBLEMS that involve real-life situations will often be written as a larger number minus a smaller number. In the last example, even if you thought that the problem should be solved by subtracting 63 from 14, that would have given you an answer of −49. It would be impossible for North to have −49 photos, since it is impossible to have a negative quantity of a real-life object. However, a word problem that simply involves numbers could have a negative answer. As always, refer to the question being asked to find the final clue of what the correct answer should be.

Example

What is fourteen less than nine?

In this example, *less than* is a backward phrase. It's also a keyword phrase that signals subtraction. To find 14 less than 9, we must start with 9 and subtract 14: $9 - 14 = -5$.

Multiplication, like addition, is commutative, so you don't have to worry about backward phrases for multiplication. Division, like subtraction, is not commutative, so we must be wary of backward phrases for division.

Example

Danielle makes five even piles out of 40 newspapers. How many newspapers are in each pile?

The keyword *each* tells us that we need to use either multiplication or division. Since Danielle is making piles from a total of 40 newspapers, each pile will have fewer than 40 newspapers, which means that we need to divide. Although five appears before 40 in the problem, it is 40 that must be divided by 5: $\frac{40}{5} = 8$. There are eight newspapers in each pile.

Example

What is 12 divided into 108?

The phrase *divided into* is a backward phrase. Twelve divided into 108 is equal to $\frac{108}{12} = 9$. The word *into* alone can also signal that the numbers of a division phrase must be reversed; *9 into 63* is $\frac{63}{9} = 7$.

CAUTION

THE PHRASE DIVIDED INTO is not the same as the phrase *divided by*. The phrase *7 divided into 28*, or sometimes simply *7 into 28*, translates into the number sentence $\frac{28}{7}$. The phrase *7 divided by 28* translates into the number sentence $\frac{7}{28}$.

DIRECTIONS: Solve each of the following word problems. Use the eight-step process, including crossing out extra information, and watch out for backward phrases.

1. Darryl and Rome each buy a video camera. Both cameras were on sale for 15% off. If Darryl paid $45 less than Rome paid, and Rome paid $340, how much did Darryl pay for his camera?
2. How many times can 14 divide 168?
3. A haystack has a diameter of 16 feet. If 13 pounds of hay are placed on top of the stack, which weighs 27 pounds, how much does the stack weigh now?
4. Mr. Miller uses five rows to seat his class of 30 students. How many students sit in each row?
5. Ryan takes 13 gumballs from a machine that contains 76 gumballs. How many gumballs are left in the machine?

PACE YOURSELF

THE WORD PROBLEMS in the last practice lap each contained numbers that appeared in the opposite order that they were used in the number sentences that solved each problem. It's usually easier to solve word problems that don't contain any backward phrases. Rewrite each word problem from the practice lap so that the numbers appear in the problem in the same order that they appear in the number sentences used to solve each problem.

MISLEADING KEYWORDS

In Chapter 1, we saw how the word *each* could signal multiplication or division. So far, we've seen the keyword phrases *more than* and *less than* signal subtraction every time. However, some word problems are written in such a way that these phrases actually signal addition.

Example

The town of Glenbrook received 12 inches of rain more than the town of Eastland over a six-week period. If Eastland received 14 inches of rain during that time, how many inches of rain did Glenbrook receive?

We'll apply the eight-step process, and watch for misleading keywords.

Read the entire word problem.
We are given the amount of rainfall Eastland received and the number of inches more than Eastland that Glenbrook received.
Identify the question being asked.
We are looking for the number of inches of rain Glenbrook received.
Underline the keywords.
The keyword phrase *more than* usually signals subtraction. However, we're given the total rainfall for Eastland, not Glenbrook. Since Glenbrook received more rainfall than Eastland, we will have to add the number of inches more to the number of inches Eastland received.
Cross out extra information and translate words into numbers.
We are told that the rainfall occurred over a six-week period, but the number 6 isn't needed to solve this problem.
List the possible operations.
Subtraction seemed like a possibility, so we will write number sentences for subtraction, but the context of the problem tells us that addition is likely the right choice.
Write number sentences for each operation.
14 + 12
14 − 12
12 − 14
Solve the number sentences and decide which answer is reasonable.
14 + 12 = 26 inches
14 − 12 = 2 inches
12 − 14 = −2 inches

Glenbrook received more rain than Eastland, so our answer must be greater than 14 inches. We're looking for an operation that increases 14, so subtraction is not the operation to use. We must add to find the total rainfall in Glenbrook.

Check your work.

We'll use subtraction to check our answer. If Glenbrook received 12 more inches than Eastland, then the number of inches Glenbrook received, 26, minus 12 should equal the number of inches Eastland received, 14: 26 – 12 = 14 inches. We have the correct operation and the answer to the question.

Example

Stuart's plane flight to China took four hours less than Cliff's flight. If Stuart's flight took five hours, how long was Cliff's flight?

Read the entire word problem.

We are given the time of Stuart's flight and how much less time his flight took than Cliff's flight took.

Identify the question being asked.

We are looking for how long Cliff's flight was.

Underline the keywords.

The keyword phrase *less than* usually signals subtraction, but we are given the time of Stuart's flight, not Cliff's flight. If we were given how long Cliff's flight was and asked to find how long Stuart's flight was, we would use subtraction. Since we are given how long Stuart's flight was, and how much less his flight was than Cliff's flight, we must add the length of his flight to the number of hours less his flight was to find how long Cliff's flight was.

Cross out extra information and translate words into numbers.

No extra information in this problem.

List the possible operations.

We're sure that addition is the operation to use to solve this problem.

Write number sentences for each operation.

Since addition is commutative, we have only one number sentence: 4 + 5

Solve the number sentences and decide which answer is reasonable.

4 + 5 = 9 hours

Stuart's flight took less time than Cliff's flight, so it makes sense that our answer for the time of Cliff's flight is greater than four hours. *Check your work.*

We'll use subtraction to check our answer. If Cliff's flight took nine hours and Stuart's flight took four hours less, the difference between 9 and 4 should be equal to the time of Stuart's flight, 5 hours: $9 - 4 = 5$ hours.

PRACTICE LAP

DIRECTIONS: Solve each of the following word problems. Use the eight-step process and watch out for misleading keywords. Some problems require addition and some require subtraction. Let the question being asked help to guide you.

6. Bria paid $129 for her bicycle. If she paid $46 less than Gavin paid for his bicycle, how much did Gavin pay for his bicycle?

7. Rori is 25 years old. If she is three years older than Bret, how old is Bret?

8. A new copy machine produces 30 copies per minute more than the old machine. If the old machine produces 28 copies per minute, how many copies per minute does the new machine produce?

9. Arizona has been a state for 76 fewer years than Arkansas has been a state. If Arkansas became a state in 1836, when did Arizona become a state?

10. The distance from Los Angeles to San Francisco is 265 miles more than the distance from Los Angeles to San Diego. If San Francisco is 389 miles from Los Angeles, how far is San Diego from Los Angeles?

OPPOSITE OPERATIONS

A word problem can be even tougher if it describes an action or operation that already occurred. Some problems will give you the result of an operation, and then ask you to find the original value. Think about how we check our work after performing an operation—after an addition problem, we use subtraction, and after a subtraction problem, we use addition. We always use the opposite operation to check our work, since we check our work by "undoing" what we did to solve the problem. If a word problem describes an operation that has already been performed, we must use the opposite of that operation to find the original value.

Example

The difference between two numbers is 34. If the smaller number is 16, what is the larger number?

Read the entire word problem.
We are given the difference between two numbers and the smaller of the two numbers.
Identify the question being asked.
We are looking for the larger of the two numbers.
Underline the keywords.
The keyword _difference_ usually signals subtraction, but the first sentence of this word problem tells us that the difference has already been found. Addition is the opposite of subtraction, so we must use addition to solve this problem.
Cross out extra information and translate words into numbers.
There is no extra information in this problem.
List the possible operations.
We must use addition to solve this problem.
Write number sentences for each operation.
Just one number sentence: 16 + 34
Solve the number sentences and decide which answer is reasonable.
16 + 34 = 50
Our answer is larger than 16, so it seems reasonable.
Check your work.
We can check our answer by finding the difference between the two numbers. We're told in the word problem that the difference is 34: 50 − 16 = 34, so we chose the right operation and answered the question correctly.

CAUTION

THE CONTEXT OF a word problem is *always* important. Even though the last example gave us a difference describing subtraction that had already been performed, it still could have been a subtraction word problem if we had been asked to find the smaller number instead of the larger number. That's why the first step in the eight-step process is to read the entire word problem, and the second step is to *identify the question being asked.*

Example

This week, 345 people purchased a subscription to a travel magazine. If 1,013 people are subscribed to the magazine now, how many people were subscribed to the magazine last week?

Read the entire word problem.
We are given the number of people currently subscribed to the magazine and the amount by which that number increased this week.
Identify the question being asked.
We are looking for the number of people who were subscribed as of last week.
Underline the keywords.
There are no keywords in this problem.
Cross out extra information and translate words into numbers.
There is no extra information in this problem.
List the possible operations.
We are told that 345 people purchased a subscription this week, which means that the number of subscriptions increased. The word *increase* usually signals addition, but we are given the new total number of subscribers, which means that this total includes the new 345 subscribers. In order to find the old number of subscribers, we must subtract 345 from the new total.
Write number sentences for each operation.
1,013 − 345

Solve the number sentences and decide which answer is reasonable.

1,013 − 345 = 668

This number is less than 1,013, which seems reasonable since we are looking for the number of subscribers before an increase of 345.

Check your work.

The number of subscribers increased by 345 to 1,013, so we can check our answer by adding 345 to it—that total should equal 1,013: 668 + 345 = 1,013 subscribers.

PRACTICE LAP

DIRECTIONS: Solve each of the following word problems. Use the eight-step process and watch out for opposite operations.

11. If the product of two numbers is 176, and one factor is 16, what is the other factor?

12. The six eighth-grade classes of Waterford Middle School take a field trip to the zoo. The students are divided into ten groups of 18. How many students took the trip to the zoo?

13. The temperature on Lake St. Jean was −9° Fahrenheit after a snowstorm dropped the temperature 15°. What was the temperature on Lake St. Jean before the storm?

14. Jordan and Jared ran through the park for two hours. If they ran 25 miles altogether, and Jordan ran 13 miles, how many miles did Jared run?

15. Brooke spent $78 at the mall. If she has $29 left in her purse, how much did she have before she went shopping?

SUMMARY

KEYWORDS CAN BE misleading—sometimes even a word like *added* can be found in a word problem that requires subtraction. By carefully reading each question and understanding the context of the problem, we can choose the right operation to solve the problem. In the next chapter, we'll look at word problems that take more than one step to solve.

ANSWERS

1. *Read the entire word problem.*

   We are given the amount of money Rome paid for his video camera and we are given the amount less that Darryl paid than Rome.

   Identify the question being asked.

   We are looking for how much Darryl paid for his camera.

   Underline the keywords.

   The keyword phrase *less than* signals subtraction.

   Cross out extra information and translate words into numbers.

   We are told that the cameras were on sale for 15% off, but we don't need this number to solve the problem—cross it out.

   List the possible operations.

   To find how much Darryl paid for his camera, we must use subtraction.

   Write number sentences for each operation.

   The phrase *less than* is a backward phrase in this problem. Although the number $45 appears first in the problem, it is $45 that must be subtracted from $340: $340 − $45.

   Solve the number sentences and decide which answer is reasonable.

   $340 − $45 = $295. Since Darryl paid less than Rome, this answer seems reasonable.

   Check your work.

   Since we used subtraction to find the answer to this problem, we must use addition to check our work. The amount Darryl paid, $295, plus the amount less he paid than Rome, $45, should equal the amount Rome paid, $340: $295 + $45 = $340.

2. *Read the entire word problem.*

   We are given the numbers 14 and 168.

   Identify the question being asked.

   How many times 14 can divide 168?

   Underline the keywords.

   The keyword *divide* tells us to use division.

   Cross out extra information and translate words into numbers.

   There is no extra information and no words that need to be translated into numbers.

   List the possible operations.

To find how many times 14 can divide 168, we must use division.

Write number sentences for each operation.

We could write 14 ÷ 168 or 168 ÷ 14, but the word *divide* alone without the word *by* tells us that we must reverse the order of the numbers given to us. We must find 168 ÷ 14.

Solve the number sentences and decide which answer is reasonable.

168 ÷ 14 = 12

Check your work.

Since we used division to find the answer to this problem, we must use multiplication to check our work. Multiply 12 by 14: 14 × 12 = 168.

3. *Read the entire word problem.*

We are given the number of pounds of hay that are added to the stack and the original weight of the stack.

Identify the question being asked.

We are looking for the new weight of the stack.

Underline the keywords.

There are no keywords in this problem.

Cross out extra information and translate words into numbers.

The diameter of the haystack is not needed to solve this problem—cross it out.

List the possible operations.

Since 13 pounds of hay was placed on the haystack, that means that the size of the haystack is increasing. We must either add or multiply to increase the original weight of the haystack.

Write number sentences for each operation.

Addition and multiplication are commutative, so we have only one number sentence for each operation. Since 13 pounds are being added (or multiplied) to 27, the number 27 should come first in our number sentences, even though both operations are commutative: 27 + 13, 27 × 13.

Solve the number sentences and decide which answer is reasonable.

27 + 13 = 40

27 × 13 = 351

An answer of 351 does not seem reasonable—that would be the total weight of 13 haystacks that each weigh 27 pounds. Addition is the right operation; 40 pounds is a reasonable answer.

Check your work.

Since we used addition to find the answer to this problem, we must use subtraction to check our work. Subtract the weight added to the stack, 13 pounds, from the new weight of the stack, 40 pounds, and that should give us the original weight of the stack, 27 pounds: $40 - 13 = 27$ pounds.

4. *Read the entire word problem.*

We are given the size of the class and the number of rows in the class.

Identify the question being asked.

We are looking for the number of students in each row.

Underline the keywords.

The keyword *each* signals multiplication or division.

Cross out extra information and translate words into numbers.

There is no extra information and no words that need to be translated into numbers.

List the possible operations.

We are given the total number of students, and we are looking for the number of students in each (one) row. When we are given the value of one and are looking for the total, we multiply; when we are given the total and we are looking for the value of one, we divide.

Write number sentences for each operation.

We could form the number sentence $5 \div 30$, but think about what the problem is telling us. Mr. Miller uses five rows to divide his class. When the word *divide* appears without the word *by*, we need to reverse the order of the numbers. The number sentence $30 \div 5$ should be used to solve the problem.

Solve the number sentences and decide which answer is reasonable.

$30 \div 5 = 6$. It is reasonable that there are six students in each of the five rows of the classroom, since there are a total of 30 students.

Check your work.

Since we used division to find the answer to this problem, we must use multiplication to check our work. Multiply the number of rows, 5, by the number of students in each row, 6, to find the total number of students in the class, 30: $5 \times 6 = 30$ students.

5. *Read the entire word problem.*

We are given the number of gumballs in a machine and the number of gumballs taken from the machine.

Identify the question being asked.

We are looking for the number of gumballs left in the machine.

Underline the keywords.

The keyword *left* signals subtraction.

Cross out extra information and translate words into numbers.

There is no extra information and no words that need to be translated into numbers.

List the possible operations.

Since gumballs have been taken from the machine, we must use an operation that decreases the total. The keyword has already told us to use subtraction.

Write number sentences for each operation.

We could form the number sentences 13 – 76 or 76 – 13. The problem tells us that Ryan takes 13 from 76, or subtracts 13 from 76. This backward phrase tells us that the number sentence to use is 76 – 13.

Solve the number sentences and decide which answer is reasonable.

76 – 13 = 63. The results of 13 – 76 would be a negative number, which would not make sense. Given that the machine contained 76 gumballs before Ryan took 13, 63 is a reasonable answer.

Check your work.

Since we used addition to find the answer to this problem, we must use subtraction to check our work. Add the number of gumballs in the machine now, 63, to the number of gumballs Ryan took, 13, and that should give us the original number of gumballs in the machine, 76: 63 + 13 = 76 gumballs.

6. *Read the entire word problem.*

We are given the amount Bria paid for her bicycle and how much less she paid than Gavin paid for his bicycle.

Identify the question being asked.

We are looking for how much Gavin paid for his bicycle.

Underline the keywords.

The keyword phrase *less than* often signals subtraction.

Cross out extra information and translate words into numbers.

There is no extra information and no words that need to be translated into numbers.

List the possible operations.

Although the phrase *less than* could mean subtraction, we are given how much Bria paid for her bicycle and the difference between how much she paid and how much Gavin paid. Since Bria paid less, we must add how much she paid to the difference between what they paid.

Write number sentences for each operation.

$129 + $46

Solve the number sentences and decide which answer is reasonable.

$129 + $46 = $175. Since we know that Bria paid less than Gavin, it makes sense that our answer, $175, is greater than $129.

Check your work.

Since we used addition to find the answer to this problem, we must use subtraction to check our work. Subtract how much Bria paid for her bicycle, $129, from how much Gavin paid for his bicycle, $175. The difference should be $46: $175 – $129 = $46.

7. *Read the entire word problem.*

We are given Rori's age and how much older she is than Bret.

Identify the question being asked.

We are looking for Bret's age.

Underline the keywords.

The keyword phrase *older than* often signals subtraction.

Cross out extra information and translate words into numbers.

There is no extra information and no words that need to be translated into numbers.

List the possible operations.

Since Rori is older than Bret and we are given Rori's age, we must subtract the difference between their ages from Rori's age to find Bret's age.

Write number sentences for each operation.

The only possible number sentence is 25 – 3, since 3 – 25 would result in a negative number.

Solve the number sentences and decide which answer is reasonable.

25 – 3 = 22. Rori is older than Bret, so we expected to find an answer that is slightly less than 25.

Check your work.

Since we used subtraction to find the answer to this problem, we must use addition to check our work. Add Bret's age to the difference between his age and Rori's age: 22 + 3 = 25, which is Rori's age.

8. *Read the entire word problem.*

   We are given the rate at which an old copy machine produces copies, and we are given how much faster the new copy machine is than the old copy machine.

   Identify the question being asked.

   We are looking for the rate at which the new copy machine produces copies.

   Underline the keywords.

   The keyword phrase *more than* often signals subtraction.

   Cross out extra information and translate words into numbers.

   There is no extra information and no words that need to be translated into numbers.

   List the possible operations.

   The keyword phrase signals subtraction, but we are given the rate of the old machine, not the rate of the new machine. Since the new machine is faster than the old one, we must add the speed of the old machine to the difference between the speeds of the two machines.

   Write number sentences for each operation.

   30 + 28

   Solve the number sentences and decide which answer is reasonable.

   30 + 28 = 58. We expected the new machine to produce more copies per minute than the old machine, so this answer is reasonable.

   Check your work.

   Since we used addition to find the answer to this problem, we must use subtraction to check our work. Subtract the speed of the new machine from the speed of the old machine. The difference should be 30 copies per minute: 58 – 28 = 30 copies per minute.

9. *Read the entire word problem.*

   We are given the year in which Arkansas became a state and how many years fewer Arizona has been a state.

   Identify the question being asked.

   We are looking for the year in which Arizona became a state.

   Underline the keywords.

   The keyword *fewer* often signals subtraction.

   Cross out extra information and translate words into numbers.

There is no extra information and no words that need to be translated into numbers.

List the possible operations.

The keyword *fewer* signals subtraction, but Arizona became a state later than Arkansas, which means that the year in which it became a state is a larger number than the year in which Arkansas became a state. We must add 76 to 1836.

Write number sentences for each operation.

76 + 1836

Solve the number sentences and decide which answer is reasonable.

76 + 1836 = 1912. Arizona became a state in 1912.

Check your work.

Since we used addition to find the answer to this problem, we must use subtraction to check our work. Subtract the year Arkansas became a state from the year Arizona became a state. That difference should be 76 years: 1912 – 1836 = 76 years.

10. *Read the entire word problem.*

We are given the distance from San Francisco to Los Angeles and how much farther that distance is than the distance from Los Angeles to San Diego.

Identify the question being asked.

We are looking for the distance from Los Angeles to San Diego.

Underline the keywords.

The keyword phrase *more than* often signals subtraction.

Cross out extra information and translate words into numbers.

There is no extra information and no words that need to be translated into numbers.

List the possible operations.

The keyword phrase *more than* signals subtraction, and since we are given the distance from San Francisco to Los Angeles, we can subtract the difference between that distance and the distance from Los Angeles to San Diego to find the distance from Los Angeles to San Diego.

Write number sentences for each operation.

389 – 265

Solve the number sentences and decide which answer is reasonable.

389 – 265 = 124 miles. San Diego is 124 miles from Los Angeles.

Check your work.

Since we used subtraction to find the answer to this problem, we must use addition to check our work. Add the distance from San Diego to Los Angeles to the difference between that distance and the distance from Los Angeles to San Francisco. The sum should be equal to the distance from Los Angeles to San Francisco: 124 + 265 = 389 miles.

11. *Read the entire word problem.*

We are given a product and one factor.

Identify the question being asked.

We are looking for the other factor.

Underline the keywords.

The keyword *product* often signals multiplication.

Cross out extra information and translate words into numbers.

There is no extra information and no words that need to be translated into numbers.

List the possible operations.

The keyword *product* signals multiplication, but since we are given a product, the multiplication has already taken place. In order to find a factor given the other factor and a product, we must divide the product by the given factor.

Write number sentences for each operation.

$\frac{176}{16}$

Solve the number sentences and decide which answer is reasonable.

$\frac{176}{16} = 11$. We can decide if our answer is reasonable by checking our work.

Check your work.

Our answer is correct if 11 multiplied by 16 is 176: $11 \times 16 = 176$, which answers the question.

12. *Read the entire word problem.*

We are given the number of groups of students and the number of students in each group.

Identify the question being asked.

We are looking for the total number of students.

Underline the keywords.

The keyword *divided* usually signals division.

Cross out extra information and translate words into numbers.

The number of eighth-grade classes, 6, is not needed to solve this problem, so cross out that number.

List the possible operations.

The keyword *divided* signals division, but the division has already occurred. The total number of students has been divided into ten groups of 18, so in order to find the total number of students, we must undo that operation and multiply 10 by 18.

Write number sentences for each operation.

10×18

Solve the number sentences and decide which answer is reasonable.

$10 \times 18 = 180$ students. Our answer seems reasonable since we would expect the total number of students to be much larger than the number of groups and the number of students in each group.

Check your work.

Since we used multiplication to find the answer to this problem, we must use division to check our work. Divide the number of students, 180, by the number of students in each group, 18. The result should be the number of groups, 10: $\frac{180}{18} = 10$ groups.

13. *Read the entire word problem.*

We are given the current temperature on the lake and the change in temperature that led to the current temperature.

Identify the question being asked.

We are looking for the temperature prior to the snowstorm.

Underline the keywords.

The keyword *dropped* often signals subtraction.

Cross out extra information and translate words into numbers.

There is no extra information and no words that need to be translated into numbers.

List the possible operations.

The keyword *dropped* signals subtraction, but we are given the temperature after this subtraction, or drop, had occurred. To find the temperature before the temperature decreased, we must add the change, 15°, to the current temperature.

Write number sentences for each operation.

$-9 + 15$

Solve the number sentences and decide which answer is reasonable.

−9 + 15 = 6°. Since the snowstorm dropped the temperature on the lake, we expected the original temperature to be higher than −9°, so our answer makes sense.

Check your work.

Since we used addition to find the answer to this problem, we must use subtraction to check our work. Subtract 15 from the original temperature, 6°, and that should give us the current temperature, −9°: 6 − 15 = −9°.

14. *Read the entire word problem.*

We are given the total number of miles Jordan and Jared ran and the number of miles Jordan ran.

Identify the question being asked.

We are looking for the number of miles Jared ran.

Underline the keywords.

The keyword *altogether* often signals addition.

Cross out extra information and translate words into numbers.

The number of hours Jordan and Jared ran is not needed to solve the problem, so cross that detail out.

List the possible operations.

The keyword signals addition, but we already have the total number of miles they ran. To find how many miles Jared ran, we must subtract the number of miles Jordan ran from the total number of miles that they both ran.

Write number sentences for each operation.

25 − 13

Solve the number sentences and decide which answer is reasonable.

25 − 13 = 12 miles. We expected our answer to be less than 25 miles, so this answer seems reasonable.

Check your work.

Since we used subtraction to find the answer to this problem, we must use addition to check our work. Add the number of miles Jordan ran, 13, to the number of miles Jared ran, 12. The total should be 25 miles: 13 + 12 = 25 miles.

15. *Read the entire word problem.*

We are given the amount of money Brooke has in her purse now and the amount of money she spent.

Identify the question being asked.

We are looking for how much money she had before she went shopping.

Underline the keywords.

The keyword *left* usually signals subtraction, as does the word *spent*.

Cross out extra information and translate words into numbers.

There is no extra information and no words that need to be translated into numbers.

List the possible operations.

The keyword *left* signals subtraction, but we are given the amount in Brooke's purse after she spent $78. To find how much money she had in her purse before she went shopping, we need to add $78 to the total left in her purse.

Write number sentences for each operation.

$29 + $78

Solve the number sentences and decide which answer is reasonable.

$29 + $78 = $107. We expected our answer to be greater than $29, so this answer makes sense.

Check your work.

Since we used addition to find the answer to this problem, we must use subtraction to check our work. Subtract the amount of money Brooke spent shopping, $78, from the amount that was in her purse before she went shopping, $107, and that should equal $29, the amount she has left: $107 – $78 = $29.

Multistep Problems

So far, the word problems we've looked at have required only one step to solve them. We used keywords or the context of the problem to decide which operation to use—but some word problems take more than one operation to solve. In this chapter, we'll look at how to find the answers to these kinds of problems.

How can you spot a word problem that requires more than one step? These problems are usually a little longer and may contain more than one keyword. If you see a keyword that signals multiplication, and then another keyword that signals subtraction, it is likely the problem will require at least two steps to solve. One of the most common types of these problems is finding change after making multiple purchases.

CAUTION

SINGLE-STEP WORD problems often contain only two numbers. Any other numbers are extra information that we don't need. However, multistep problems often contain more than two numbers—and we need all of them. When deciding whether information is extra, be sure to think about all of the steps needed to solve the problem.

Example

Alana buys three pairs of shoes, each for $49. If she gives the cashier $150, how much change will she receive?

Read the entire word problem.

We are given the number of shoes Alana buys, the price of one pair of shoes, and how much Alana gives the cashier.

Identify the question being asked.

We are looking for how much change she receives.

Underline the keywords.

The keyword *each* can signal multiplication or division, and the word *change* often signals subtraction.

Cross out extra information and translate words into numbers.

This problem contains three numbers, but we will need all three numbers to solve the problem. The number of pairs of shoes and the price of each pair are needed to determine the total price. The amount Alana gave the cashier will be needed to find her change. There is no extra information in this problem.

List the possible operations.

We are given the price of one pair of shoes, and we are looking for the price of three pairs of shoes. When we are given the value of one and we're looking for the total value of more than one, we need to use multiplication. In order to find change, we need to subtract a total price from an amount of money paid. To solve this problem, we will need to multiply first, and then subtract.

Write number sentences for each operation.

The total Alana must pay is equal to three times the price of one pair of shoes: 49×3. We can't form our subtraction sentence yet because we need the result of that multiplication sentence.

Solve the number sentences and decide which answer is reasonable.

$49 \times 3 = 147

Since the price of one pair of shoes is $49, $147 seems like a reasonable price for three pairs of shoes. However, this is not our answer, since we are looking for how much change Alana receives after she gives the cashier $150. This shows why it is important to identify the question being asked. We must return to the previous step and write another number sentence.

Write number sentences for each operation.

Alana gives the cashier $150 for shoes that cost a total of $147. Her change will be equal to the difference between these two numbers: $150 – $147. This is our only number sentence; it would not make sense to subtract $150 from $147.

Solve the number sentences and decide which answer is reasonable.

$150 – $147 = $3. Alana receives $3 in change. Since she paid only slightly more than the total cost, this answer makes sense.

Check your work.

Since solving the problem took two steps, checking our work will take two steps. Add the amount of change Alana received for her total bill. This should equal how much she gave the cashier: $3 + $147 = $150. Divide Alana's total bill by 3. This should equal the cost of one pair of shoes: $\frac{\$147}{3}$ = $49. Our answer is correct.

To check the answer to a multistep word problem, perform the opposite operations you used to solve the problem in the opposite order. In the last example, we multiplied and then subtracted. To check our work, we added and then divided. Since checking our work is like "undoing" our work, we need to not only perform the opposite operations, but do them in reverse order.

Steps 5 and 6 of the process were repeated in the last example because we needed two operations to solve the problem. We will perform those two steps for every operation needed to solve a word problem. If a problem requires four operations, then we'll perform those steps four times.

Example

Tiffani received $15 from each of her uncles on her birthday, plus $25 from her parents. If Tiffani has five uncles, how much money did she receive altogether?

Read the entire word problem.

We are given how much money Tiffani receives for each of her uncles, the number of uncles, and how much money she receives from her parents.

Identify the question being asked.

We're looking for how much money she receives altogether.

Underline the keywords.

The keyword *each* can signal multiplication or division, and the keywords *plus* and *altogether* signal addition.

Cross out extra information and translate words into numbers.

Just as in the last example, this problem contains three numbers, but we will need all three to find our answer. There is no extra information here.

List the possible operations.

Since Tiffani receives $15 from each of her uncles, and we are looking for how much she receives from all five of her uncles, we must multiply. She also receives $25 from her parents, so we must add that amount to the total she receives from her uncles.

Write number sentences for each operation.

Multiply the amount Tiffani receives from each uncle by the number of uncles: $15 × 5. We can't form our addition sentence yet because we need the result of that multiplication sentence.

Solve the number sentences and decide which answer is reasonable.

$15 × 5 = $75

Seventy-five dollars is a reasonable number for Tiffani to have received from her uncles, given that she has five of them and each gave $15. Now that we have the total she received from them, we can add to it the amount she received from her parents.

Write number sentences for each operation.

Tiffani received $75 from her uncles and $25 from her parents: $75 + $25.

Solve the number sentences and decide which answer is reasonable.

$75 + $25 = $100

Check your work.

Checking our work will require two steps. Since we used multiplication and addition, in that order, to solve the problem, we will use subtraction and division, in that order, to check our answer. First, take the total Tiffani received and subtract from it how much she received from her parents. This should equal how much she received from her uncles: $100 – $25 = $75. Now divide that amount by the number of uncles. This should equal how much she received from each uncle: $75 ÷ 5 = $15.

Each of the two examples we've looked at required two different operations to find our answer. It is also possible for a problem to require the same operation to be performed more than once.

Example

Don owns three farms. Each farm has five chicken coops, and each chicken coop holds 25 chickens. If each chicken lays six eggs per week, how many eggs will Don have at the end of one week?

This problem contains an awful lot of information, but we will use the same process to solve it, repeating steps 5 and 6 until we have solved the problem.

Read the entire word problem.
We are given the number of eggs laid by a chicken in a week, the number of chickens in a chicken coop, the number of chicken coops on a farm, and the number of farms.
Identify the question being asked.
We are looking for the total number of eggs laid in one week.
Underline the keywords.
The keyword *each* can signal multiplication or division, and this keyword appears in the word problem three times!
Cross out extra information and translate words into numbers.
The numbers 3, 5, 25, and 6 will all be needed to solve the problem. In fact, even the number 1 (the number of weeks) is important. Multiplying by 1 won't change our answer, but if we had been asked to look for the number of eggs laid in two weeks, then that number would be very important.
List the possible operations.
We are given the number of farms and the number of chicken coops on each farm. We will need multiplication to find the total number of chicken coops. We are given the number of chickens in each coop, so we'll need multiplication again to find the total number of chickens.

Once we have that number, we'll need multiplication one more time to find the number of eggs the chickens laid in a week. If we needed to find how many eggs the chickens laid in more than one week, we would need to multiply yet again!

Write number sentences for each operation.

We'll form one multiplication sentence at a time, since we'll need our answer to each in order to form the next sentence. First, find the number of chicken coops on the three farms: 3×5.

Solve the number sentences and decide which answer is reasonable.

$3 \times 5 = 15$

Now that we have the number of chicken coops, we can return to step 5 and find the total number of chickens.

Write number sentences for each operation.

There are 15 chicken coops and 25 chickens in each coop, so the total number of chickens is equal to 15×25.

Solve the number sentences and decide which answer is reasonable.

$15 \times 25 = 375$

Return to step 5 and use the total number of chickens to find the total number of eggs laid.

Write number sentences for each operation.

There are 375 chickens and each lays six eggs per week, so the total number of eggs laid is equal to 375×6.

Solve the number sentences and decide which answer is reasonable.

$375 \times 6 = 2,250$

Since we are looking for the number of eggs laid in one week, we have our answer.

Check your work.

We multiplied three times to find our answer, so we will check our work by dividing three times. Divide the total number of eggs by the number of eggs laid by each chicken. This should equal the number of chickens: $2,250 \div 6 = 375$. Divide the number of chickens by the number of chickens in each coop. This should equal the number of chicken coops: $375 \div 25 = 15$. Finally, divide the number of chicken coops by the number of chicken coops on each farm. This should equal the number of farms: $15 \div 3 = 5$. Our answer is correct.

1. Rikin has $328 in his bank account. He deposits $103, and then takes $66 out of the account. How much money is in Rikin's account now?

2. Nicki bakes four dozen cupcakes. She eats three cupcakes and shares the rest with nine friends. How many cupcakes does each friend eat?

3. Alison is decorating for a party. She buys six packages of balloons, each of which contains 12 balloons. Alison places all of the balloons around a room, but four of them pop. How many balloons does Alison have left?

4. A truck delivers newspapers to 24 newsstands. Each newsstand receives 50 newspapers. If each newspaper costs $0.50, and all of the newspapers are sold, how much money will the newsstands collect?

5. Brittany buys 500 beads. She needs 65 beads to make a necklace. If she makes six necklaces, how many beads will she have left?

6. One package of paper contains 140 sheets. Jason's report is six pages long, and he must make one copy for each student in the ninth grade. If there are 210 students in the ninth grade, how many packages of paper does Jason need?

7. Andrew has baseball practice three times a week. Each practice is two hours long, and Andrew throws 50 pitches per hour. If he uses a new baseball after every ten pitches, how many baseballs does Andrew use in four weeks?

8. A pizzeria charges $1.85 for a slice of pizza and $1.25 for a drink. Ted buys two slices of pizza and a drink and pays with a $5 bill. How much change does he receive?

9. A cooler holds 280 ounces of water. Michelle uses the cooler to fill 18 eight-ounce cups of water, and then fills a 36-ounce pitcher. How much water is left in the cooler?

10. Joy's dance class is two hours long and meets four times per week. Sarah's dance class is three hours long and meets three times per week. How many more hours does Sarah dance than Joy in a year?

ORDER OF OPERATIONS

Whenever we must use more than one operation to solve a problem, we need to be careful about the order in which we do those operations. Sometimes the order of operations is easy to spot based on the context of the problem, but when a word problem is strictly about numbers, it can be more difficult. For problems like this, we must read carefully and break the problem into pieces.

Example

What is three less than twice four?

We can use the step-by-step process to solve problems like this, but first, let's focus on each word of the problem. We are asked to find three less than twice four. Before we can find three less than twice four, we must first find twice four. Twice four is the same as two times four. Once we have found two times four, we can find three less than that by subtracting three. This problem, written numerically, is $(2 \times 4) - 3$. Now that we know what the problem is asking us, we don't need the eight-step process—it's not a word problem anymore. $(2 \times 4) - 3 = 8 - 3 = 5$. Three less than twice four is five.

> ## CAUTION
>
> **THE ORDER OF OPERATIONS** is Parentheses, Exponents, Multiplication, Division, Addition, Subtraction. However, the order in which to perform operations in a word problem depends on how the operations are described. A multistep problem may require multiplication and addition, but you may have to add first. When breaking a word problem down into a numeric problem, be sure to use parentheses to help you group operations together. Remember, operations in parentheses, no matter what they are, come before all other operations.

Example

Find four times the sum of six and ten.

This problem asks us to find four times a sum. In order to multiply that sum by four, we must first find what that sum is. Our first step is to add six and ten. Then, we will multiply by four. This problem, written numerically, is $4 \times (6 + 10)$. It is important to put the addition of 6 and 10 in parentheses, because $4 \times (6 + 10)$ is not the same as $4 \times 6 + 10$. Putting parentheses around $6 + 10$ reminds us that this sum must be found before we can multiply: $4 \times (6 + 10) = 4 \times 16 = 64$.

Example

What is the product of nine less than twenty and three more than fifteen?

The problem begins by asking us to find a product. We will have to multiply two numbers that we don't have yet, so multiplication will be the last operation we perform:

$$(\ ) \times (\ )$$

The next part of the word problem is nine less than twenty. We can find nine less than twenty by subtracting 9 from 20: 20 − 9. The result of this subtraction will be the first number in our multiplication sentence:

$$(20 - 9) \times (\ )$$

The last part of the word problem asks us to find three more than fifteen. The keyword phrase *more than* often signals subtraction, but to find the number that is three more than 15, we must add 3 to 15: 15 + 3. Now we can place this sum into our number sentence:

$$(20 - 9) \times (15 + 3)$$

We've set up our number sentence and we're ready to solve. 20 − 9 = 11 and 15 + 3 = 18, so the problem becomes 11 × 18 = 198.

INSIDE TRACK

WHEN A WORD problem is made up of only numbers, separate the problem into single operations and put parentheses around each so that you solve the operations in the right order. Save the eight-step process for more complicated word problems!

PRACTICE LAP

11. Find the quotient of ninety-six and three less than eleven.

12. What is five more than the product of six and eight?

13. What number is sixteen times the sum of two and ten?

14. Find the difference between twenty and twice seven.

15. What is nineteen multiplied by twelve divided by four?

16. Find the sum of six dozen and six less than twelve.

17. What is the product of seven fewer than thirteen and five squared?

18. What is the product of four and six less than ten, divided by two?

19. Find the difference between eight more than seventeen and eight less than seventeen.

20. What is the product of the difference between sixteen and nine and the sum of eleven and twice five?

PACE YOURSELF

WHY ARE PARENTHESES so important when setting up a multistep problem? Take the numbers 2, 4, 6, 8, and 10. Use these numbers to form a number sentence that includes addition, subtraction, multiplication, and division, and then find the answer to that number sentence. Now, place parentheses around the addition part of your number sentence, and then place another set of parentheses around the subtraction part of your number sentence. Did your answer change? Why?

SUMMARY

IN THIS CHAPTER, we learned how to repeat parts of the eight-step process to solve multistep word problems and how to convert word problems that are made up of only numbers into number sentences with parentheses. We've also shown how to use keywords, pictures, tables, diagrams, and number sentences to solve single-step and multistep problems, and what pitfalls to avoid when working with word problems. Our problem-solving foundation is set—we're ready to look at word problems by subject. The next six chapters each focus on word problems in one specific area of math.

ANSWERS

1. *Read the entire word problem.*

   We are given how much Rikin has in his account, how much he deposits, and how much he takes out.

   Identify the question being asked.

   We are looking for how much money he has in his account now.

   Underline the keywords.

   There are no keywords in this problem, but a deposit increases the amount of money in an account, and a withdrawal from an account decreases that total.

   Cross out extra information and translate words into numbers.

   There is no extra information and no words that need to be translated into numbers.

   List the possible operations.

   Since a deposit increases the amount in Rikin's account, we must add $103 to his account. When he takes $66 out of the account, we must subtract $66 from his total.

   Write number sentences for each operation.

   Write the addition number sentence first. We will use this sum in our subtraction sentence: $328 + $103.

   Solve the number sentences and decide which answer is reasonable.

   $328 + $103 = $431. We expected our answer to be greater than $328, so this answer is reasonable.

   Write number sentences for each operation.

   Now write the subtraction number sentence using the sum we just found: $431 – $66.

   Solve the number sentences and decide which answer is reasonable.

   $431 – $66 = $365

   Check your work.

   We added and then subtracted to find our answer, so we will check our work by adding (the opposite of subtracting) and then subtracting (the opposite of adding). Adding the amount Rikin took out of his account to his final total should give us the amount in his account after his deposit: $365 + $66 = $431. Subtracting the deposit from this total should give us the original amount of money in Rikin's account: $431 − $103 = $328.

2. *Read the entire word problem.*

We are given how many cupcakes Nicki bakes, how many she eats, and how many friends with whom she shares her cupcakes.

Identify the question being asked.

We are looking for how many cupcakes each friend eats.

Underline the keywords.

The keyword *share* signals division.

Cross out extra information and translate words into numbers.

The word *dozen* means "12," and the phrase *four dozen* means "four times twelve": 4 × 12 = 48. Replace the phrase *four dozen* with "48."

List the possible operations.

Eating cupcakes signals a decrease in the total number of cupcakes, so we will need to subtract the number of cupcakes Nicki eats from the total, 48. Then, she shares the remainder with her nine friends, so we will need to divide the result of our subtraction by 9.

Write number sentences for each operation.

Write the subtraction number sentence first. We will use this difference in our division sentence: 48 − 3.

Solve the number sentences and decide which answer is reasonable.

48 − 3 = 45. 45 is slightly less than 48, so our answer is reasonable.

Write number sentences for each operation.

Now write the division number sentence using the difference we just found: 45 ÷ 9.

Solve the number sentences and decide which answer is reasonable.

45 ÷ 9 = 5

Check your work.

We subtracted and then divided to find our answer, so we will check our work by multiplying and then adding. Multiply the number of friends, 9, by the number of cupcakes each friend eats: 9 × 5 = 45. Add to this product the number of cupcakes Nicki eats, 3. This should give us the original number of cupcakes Nicki baked, 48: 45 + 3 = 48 cupcakes.

3. *Read the entire word problem.*

We are given how many packages of balloons Alison buys, the number of balloons in each package, and the number of balloons that pop.

Identify the question being asked.

We are looking for how many balloons are left.

Underline the keywords.

The keyword *each* can signal multiplication or division, and the keyword *left* signals subtraction.

Cross out extra information and translate words into numbers.

There is no extra information and no words that need to be translated into numbers.

List the possible operations.

Alison buys six packages of balloons, and each package contains 12 balloons. Since one package contains 12 balloons, we will need to multiply to find the number of balloons in six packages. We will subtract from that product the number of balloons that pop.

Write number sentences for each operation.

6×12

Solve the number sentences and decide which answer is reasonable.

$6 \times 12 = 72$

Write number sentences for each operation.

Now write the subtraction number sentence using the product we just found: $72 - 4$.

Solve the number sentences and decide which answer is reasonable.

$72 - 4 = 68$

Check your work.

We multiplied and then subtracted to find our answer, so we will check our work by adding and then dividing. Add the number of balloons that were left to the number of balloons that popped: $4 + 68 = 72$. Divide the number of balloons, 72, by the number of packages, 6, and this should give us the number of balloons in each package, 12: $72 \div 6 = 12$ balloons.

4. *Read the entire word problem.*

We are given the number of newsstands, the number of newspapers at each newsstand, and the price of each newspaper.

Identify the question being asked.

We are looking for the total money collected.

Underline the keywords.

The keyword *each* can signal multiplication or division, and it appears twice in the problem.

Cross out extra information and translate words into numbers.

There is no extra information and no words that need to be translated into numbers.

List the possible operations.

Since one newsstand receives 50 newspapers, we will need to multiply the number of newsstands by 50 to find the total number of newspapers delivered. Once we have that number, we can multiply it by the price of one newspaper.

Write number sentences for each operation.

24 × 50

Solve the number sentences and decide which answer is reasonable.

24 × 50 = 1,200

Write number sentences for each operation.

There are 1,200 newspapers and each costs $0.50. The total money collected is equal to the product of 1,200 and $0.50.

Solve the number sentences and decide which answer is reasonable.

1,200 × $0.50 = $600

Check your work.

We multiplied twice to find our answer, so we will check our work by dividing twice. Divide the total money collected by the price of one newspaper to give us the number of newspapers sold: $600 ÷ $0.50 = 1,200. Divide the number of newspapers sold by the number of newspapers delivered to each newsstand. This should give us the number of newsstands: 1,200 ÷ 50 = 24 newsstands.

5. *Read the entire word problem.*

We are given the number of beads Brittany buys, the number of beads in a necklace, and the number of necklaces Brittany makes.

Identify the question being asked.

We are looking for the number of beads she has left.

Underline the keywords.

The keyword *left* signals subtraction.

Cross out extra information and translate words into numbers.

There is no extra information and no words that need to be translated into numbers.

List the possible operations.

First, find how many beads were used to make the six necklaces. Since we have the number of beads needed to make one necklace, we can multiply to find the number of beads needed to make six necklaces.

Write number sentences for each operation.

6×65

Solve the number sentences and decide which answer is reasonable.

$6 \times 65 = 390$

Write number sentences for each operation.

Since Brittany began with 500 beads, subtract the 390 beads she used to make the necklaces to find how many beads she has left: $500 - 390$.

Solve the number sentences and decide which answer is reasonable.

$500 - 390 = 110$ beads

Check your work.

We multiplied and subtracted to find our answer, so we will add and divide to check our work. Add the number of beads remaining to the number of beads used, and this should equal the original number of beads: $110 + 390 = 500$. Divide by 6 the number of beads used to make the necklaces to give us the number of beads needed to make one necklace: $390 \div 6 = 65$ beads.

6. *Read the entire word problem.*

We are given the number of sheets of paper in a package, the length of Jason's report, and the number of students in the ninth grade.

Identify the question being asked.

We are looking for the number of packages of paper Jason needs.

Underline the keywords.

The keyword *each* can signal multiplication or division.

Cross out extra information and translate words into numbers.

There is no extra information and no words that need to be translated into numbers.

List the possible operations.

If one copy of Jason's report is six pages, and he needs to make 210 copies of his report, we need to multiply to find the total number of pages Jason needs.

Write number sentences for each operation.

6×210

Solve the number sentences and decide which answer is reasonable.

$6 \times 210 = 1,260$

Write number sentences for each operation.

One package of paper contains 140 sheets, so the number of packages of paper that Jason needs is equal to $\frac{1,260}{140}$.

Solve the number sentences and decide which answer is reasonable.

$\frac{1,260}{140} = 9$ packages

Check your work.

We multiplied and divided to find our answer, so we will divide and multiply to check our work. Multiply the number of packages, 9, by the number of sheets in a package, 140. This should equal the total number of sheets Jason needs: $9 \times 140 = 1,260$. Divide this product by the number of students in the ninth grade, 210, and this should equal the length of Jason's report: $\frac{1,260}{210} = 6$ pages.

7. *Read the entire word problem.*

We are given the number of times Andrew practices each week, the length of each practice, the number of pitches he throws, and how often he needs a new baseball.

Identify the question being asked.

We are looking for the number of baseballs Andrew uses in four weeks.

Underline the keywords.

The keyword *times* signals multiplication, and the keywords *each*, *per*, and *every* can signal multiplication or division.

Cross out extra information and translate words into numbers.

There is no extra information and no words that need to be translated into numbers.

List the possible operations.

Begin by using the number of practices Andrew has each week and the length of each practice to find how many hours he practices each week. Since we know the length of one practice is two hours, we must multiply to find the length of three practices.

Write number sentences for each operation.

2×3

Solve the number sentences and decide which answer is reasonable.

$2 \times 3 = 6$

Write number sentences for each operation.

Andrew throws 50 pitchers per hour, so we must multiply again to find the number of pitches he throws in six hours: 50×6.

Solve the number sentences and decide which answer is reasonable.

$50 \times 6 = 300$ pitches

Write number sentences for each operation.

Andrew uses a new baseball after every ten pitches, so to find how many baseballs he needs to throw 300 pitches, we must divide: $300 \div 10$.

Solve the number sentences and decide which answer is reasonable.

$300 \div 10 = 30$ pitches

Check your work.

We multiplied twice and then divided to find our answer, so we will multiply once and divide twice to check our work. Multiply the number of baseballs Andrew used by the number of pitches he throws with one baseball: $30 \times 10 = 300$ pitches. Andrew throws 50 pitches per hour, so divide 300 by 50 to find the number of hours Andrew practices: $300 \div 50 = 6$. Since each practice is two hours long, 6 divided by 2 should equal the number of practices Andrew has each week: $6 \div 2 = 3$ practices.

8. *Read the entire word problem.*

We are given the price of a slice of pizza, the price of a drink, the number of slices and the number of drinks Ted buys, and how much he pays.

Identify the question being asked.

We are looking for the change he receives.

Underline the keywords.

There are no keywords in this problem, but the word *change* often signals subtraction.

Cross out extra information and translate words into numbers.

There is no extra information and no words that need to be translated into numbers.

List the possible operations.

First, we must find how much Ted spent. Begin by multiplying the number of slices Ted bought by the price of each slice.

Write number sentences for each operation.

$2 \times \$1.85$

Solve the number sentences and decide which answer is reasonable.

$2 \times \$1.85 = \$3.70$

Write number sentences for each operation.

Next, find the total amount Ted spent by adding the cost of the drink: $3.70 + $1.25.

Solve the number sentences and decide which answer is reasonable.

$3.70 + $1.25 = $4.95. Now that we have Ted's total bill, we can find how much change he received by subtracting that total from the amount Ted paid, $5.

Write number sentences for each operation.

$5.00 – $4.95

Solve the number sentences and decide which answer is reasonable.

$5.00 – $4.95 = $0.05

Check your work.

We multiplied, added, and subtracted to find our answer, so we will add, subtract, and divide to check our work. Add the change Ted received to his total bill, and this sum should equal the amount Ted paid: $0.05 + $4.95 = $5.00. Subtract from Ted's bill the cost of the drink, and this difference should equal how much Ted paid for the two slices of pizza: $4.95 – $1.25 = $3.70. Finally, divide that total by two, and the answer should equal the price of one slice of pizza: $3.70 ÷ 2 = $1.85.

9. *Read the entire word problem.*

We are given the number of ounces in the cooler, the number of cups of water filled, the size of each cup, and the number of ounces used to fill a pitcher.

Identify the question being asked.

We are looking for the amount of water left in the cooler.

Underline the keywords.

The keyword *left* signals subtraction.

Cross out extra information and translate words into numbers.

There is no extra information and no words that need to be translated into numbers.

List the possible operations.

In order to find how much water is left in the cooler, we must first figure out how much water was poured into the cups. Since there are 18 cups, and one cup holds 8 ounces, we must multiply 18 by 8 to find how many ounces were used to fill the cups.

Write number sentences for each operation.

18×8

Solve the number sentences and decide which answer is reasonable.

$18 \times 8 = 144$

Write number sentences for each operation.

Subtract 144 ounces from the total volume in the cooler, 280 ounces.

Solve the number sentences and decide which answer is reasonable.

$280 - 144 = 136$ ounces. We don't have our answer yet because this doesn't fully answer the question. We still need to subtract the 36 ounces Michelle poured into the pitcher.

Write number sentences for each operation.

$136 - 36$

Solve the number sentences and decide which answer is reasonable.

$136 - 36 = 100$ ounces

Check your work.

We multiplied and then subtracted twice to find our answer. We will add twice and divide to check our work. Add the number of ounces poured into the pitcher to the final volume of the cooler: $100 + 36 = 136$. Add the number of ounces in the 18 cups to this sum, and that sum should give us the original amount of water in the cooler: $136 + 144 = 280$ ounces. We can check that our multiplication was correct by dividing the total volume of the cups, 144 ounces, by the number of ounces in each cup, 8, and that quotient should give us the number of cups, 18: $144 \div 8 = 18$ cups.

10. *Read the entire word problem.*

We are given length of each girl's dance class and the number of times per week each class meets.

Identify the question being asked.

We are looking for how many more hours Sarah dances than Joy dances.

Underline the keywords.

The keyword *per* appears twice in the problem and can signal multiplication or division. The keyword phrase *more than* often signals subtraction.

Cross out extra information and translate words into numbers.

A year is made up of 52 weeks. After finding how many hours each girl dances in a week, that product must be multiplied by 52 to find how many hours each girl dances in a year.

List the possible operations.

First, find the number of hours Joy dances in a week by multiplying the number of hours in one class by the number of classes in a week.

Write number sentences for each operation.

2×4

Solve the number sentences and decide which answer is reasonable.

$2 \times 4 = 8$

Write number sentences for each operation.

Now find the number of hours Sarah dances each week using a similar number sentence.

3×3

Solve the number sentences and decide which answer is reasonable.

$3 \times 3 = 9$. We have two options now. We can either subtract to find the difference in the number of hours per week, and then multiply that difference by 52 to find the difference in hours for the year, or we can multiply by 52 the number of hours per week danced by each girl, and then subtract. The math will be a little easier if we subtract first, but either option will work. Subtract the number of hours per week that Joy dances from the number of hours per week that Sarah dances.

Write number sentences for each operation.

$9 - 8$

Solve the number sentences and decide which answer is reasonable.

$9 - 8 = 1$

Write number sentences for each operation.

Multiply the difference by 52 to find how many more hours Sarah dances than Joy dances in a year: 1×52.

Solve the number sentences and decide which answer is reasonable.

$1 \times 52 = 52$ hours

Check your work.

Divide the number of hours that Sarah dances more than Joy in a year by the number of weeks in a year: $52 \div 52 = 1$ hour. This is the number of hours more that Sarah dances than Joy dances each week. Add this to the number of hours Joy dances each week: $1 + 8 = 9$. Divide 9 by the number of times Sarah dances each week, and this should equal the number of hours of each dance class: $9 \div 3 = 3$ hours.

11. We are looking for the quotient of 96 and a number: $96 \div (\ )$. That number is three less than eleven, which is $11 - 3$. $96 \div (11 - 3) = 96 \div 8 = 12$.

12. We are looking for 5 more than a number: $5 + (\ )$. That number is the product of six and eight, which is 6×8. $5 + (6 \times 8) = 5 + 48 = 53$.

13. We are looking for 16 times a number: $16 \times (\ )$. That number is the sum of two and ten, which is $2 + 10$. $16 \times (2 + 10) = 16 \times 12 = 192$.

14. We are looking for the difference between 20 and a number: $20 - (\ )$. That number is twice seven, which is 2×7; $20 - (2 \times 7) = 20 - 14 = 6$.

15. We can either multiply 19 by 12 and then divide by 4, or we can divide 12 by 4 and then multiply by 19. In this problem, the order does not matter. Let's start by dividing, since it will make the numbers easier to work with. $19 \times (12 \div 4) = 19 \times 3 = 57$.

16. We are looking for the sum of two numbers: $(\ ) + (\ )$. The first number is six dozen, which is 6×12. The second number is six less than twelve, which is $12 - 6$. The problem is now $(6 \times 12) + (12 - 6) = 72 + 6 = 78$.

17. We are looking for the product of two numbers: $(\ ) \times (\ )$. The first number is seven fewer than thirteen, which is $13 - 7$. The second number is five squared, which is 5^2. The problem is now $(13 - 7) \times (5^2) = 6 \times 25 = 150$.

18. We are looking for the product of two numbers divided by two: $[(\ ) \times (\ )] \div 2$. The first number is four. The second number is six less than ten, or $10 - 6$. The problem is now $[(4) \times (10 - 6)] \div 2 = (4 \times 4) \div 2 = 16 \div 2 = 8$.

19. We are looking for the difference between two numbers: $(\ ) - (\ )$. The first number is eight more than seventeen, which is $17 + 8$. The second number is eight less than seventeen, which is $17 - 8$. The problem is now $(17 + 8) - (17 - 8) = 25 - 9 = 16$.

20. We are looking for the product of two numbers: $(\ ) \times (\ )$. The first number is the difference between sixteen and nine, which is $16 - 9$. The second number is the sum of two numbers. The first is 11 and the second is twice five, or 2×5. The problem is now $(16 - 9) \times [11 + (2 \times 5)] = 7 \times (11 + 10) = 7 \times 21 = 147$.

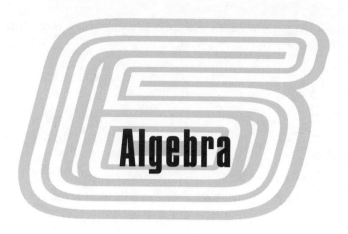

Algebra

Now that we have a set of strategies for solving word problems, the next six chapters will review how to solve word problems on six different math topics. We'll start with algebra—although you may find these problems look very much like the multistep problems you worked with at the end of Chapter 5.

ALGEBRAIC EXPRESSIONS: SUBJECT REVIEW

An **algebraic expression** can be a single **term** that contains a **variable**, such as $3x$, or many terms, such as $2x^3 + 5x^2 - 7$.

> A **term** is a variable, a number, or a variable *and* a number that is multiplied, divided, or raised to an exponent. Terms can be added or subtracted.
> A **variable** is a letter or symbol in an equation or an expression that holds the place of a number.
> An **algebraic expression** is a single term or multiple terms on which one or more operations are performed and in which at least one variable is present.

A word problem can describe an algebraic expression in words and ask you to write that expression in numbers or variables, the way you normally see it. First, let's practice going in the opposite direction: writing algebraic expressions in words.

Example

What is $3x$ in words?

When a number, or constant, appears right next to a variable, that constant is multiplied by the variable. How can we describe the variable x? It is an unknown number, or simply "a number." $3x$ is "three times a number."

A **constant** is a real number, such as 8 or –1, and not a variable.

Example

Describe $x + 7$ in words.

We still call x "a number," so $x + 7$ is "seven more than a number" or "a number plus seven."

Describing a single operation may not be hard, but when an expression contains multiple operations, we must be careful how we write the expression in words, in order to preserve the correct order of operations and the correct interpretation.

Example

What is $-7a + 5$ in words?

This expression contains two terms, $-7a$ and 5. However, –7 is multiplied by a before 5 is added, so we must be sure to keep the order of operations clear when we describe this expression. $-7a$ is "the product of negative seven and a number," so $-7a + 5$ is "the product of negative seven and a number, added to 5." If we had written "the product of negative seven and the sum of a number and 5," we would have described the expression $-7(a + 5)$, which is equal to $-7a - 35$, not $-7a + 5$, so the correct interpretation is just as important as the order of operations.

Now that we are comfortable turning algebraic expressions into words, we're ready to handle algebraic word problems. Given an algebraic expression in words, write the algebraic expression.

WHEN CONVERTING A phrase or a sentence into an algebraic expression, remember to check for keywords and backward phrases, just as you would when working with a word problem that doesn't involve algebra.

Example

What is three less than a number?

This question is just like the questions we saw at the end of Chapter 5, only instead of writing a word problem solving an expression made up of only constants, we will be writing an expression that is a mix of constants and variables. Just as with those problems, once we've converted the word problem into an algebra problem, we no longer need the eight-step process, because we no longer have a word problem.

Whenever you see a quantity in a word problem described as "a number" or "an unknown," you know it's an algebra problem, and you will need to use a variable to represent that number. Most people like to use "x" to represent unknown numbers, so we will do the same, but you could use any letter. If we use x to represent "a number" in this word problem, then "three less than a number" becomes "three less than x." Rewrite any number words as numbers. The expression is now "3 less than x." As we learned in Chapter 1, the words *less than* signal subtraction, and as we learned in Chapter 4, these words can also be a backward phrase: *3 less than* x is "$x - 3$."

Example

Find seven divided by the sum of nine and a number.

The keywords *divided* and *sum* tell us that we will need to use division and addition. Convert the words that are numbers to real numbers, and rewrite *a number* as "x": Find 7 divided by the sum of 9 and x. Since 7 is divided by a sum, we must write the sum before we can divide. The sum of 9 and x is $9 + x$. The number 7 divided by that sum is $\frac{7}{(9 + x)}$.

CAUTION

WE MAY NEED to add parentheses to an algebraic expression to preserve the correct order of operations. In the last example, if we had written $\frac{7}{9} + x$, our answer would have been incorrect. Division is performed before addition in the order of operations, so $\frac{7}{9} + x$ is equal to "the sum of seven divided by nine and a number" and not "seven divided by the sum of nine and a number." Always place parentheses around the operation you want to be performed first. Even if that operation would have been performed first anyway, it would not be incorrect to place the parentheses. If we wanted to write the expression "the sum of seven divided by nine and a number," we could write either $\frac{7}{9} + x$ or $\left(\frac{7}{9}\right) + x$, and both expressions would be correct.

PRACTICE LAP

DIRECTIONS: Write each of the following phrases as an algebraic expression.

1. eight more than a number
2. a number divided by six
3. eleven times the sum of a number and five
4. the quotient of a number and the difference between nineteen and two
5. twice the product of two different numbers
6. one more than the difference between six and a number
7. the square of a number, minus five times that same number
8. ten fewer than fifteen times the cube of a number

PACE YOURSELF

WRITE AN ALGEBRAIC expression to represent what your height will be in five years. What numbers or variables are in your expression? How could you evaluate your expression five years from now?

ALGEBRAIC EQUATIONS: SUBJECT REVIEW

Just as we wrote algebraic expressions from word problems, we can write algebraic equations from word problems and solve them for the unknown number.

An **algebraic equation** is an algebraic expression with an equals sign, such as $3x = 9$.

We can solve an algebraic equation by isolating the variable on one side of the equation using addition, subtraction, multiplication, division, or exponents. For example, the equation $2x - 6 = 12$ can be solved by adding 6 to both sides of the equation and dividing by 2. This will isolate x on the left side of the equation: $2x - 6 + 6 = 12 + 6$, $2x = 18$, $\frac{2x}{2} = \frac{18}{2}$, $x = 9$. We can convert word problems to algebraic equations, and then solve them for x.

The difference between an algebraic expression word problem and an algebraic equation word problem is that an algebraic expression problem will ask you to write an expression with an unknown number, and an algebraic equation problem will ask you to find the value of an unknown number.

Example

If nine more than a number is thirteen, what is the number?

We begin to solve this kind of problem just as we began algebraic expression word problems. Write *nine* as "9," write *a number* as "x," and write *thirteen* as "13." However, we have a new keyword: *is*. The word *is* represents the equals sign. When you see the word *is*, the word *equals*, or the phrase *is equal to*, include the equals sign. *Nine more than a number is thirteen* becomes "9 more than $x = 13$." The phrase *more than* signals addition, so we now have

$x + 9 = 13$. To solve for x, subtract 9 from both sides of the equation: $x + 9 - 9 = 13 - 9$, $x = 4$.

Sometimes, an equation can have an unknown on both sides of the equals sign.

Example

> If three times a number is equal to ten more than the number, what is the number?

Again, begin by converting words to numbers and unknowns to variables. *Three* becomes "3," *a number* becomes "x," and *ten* becomes "10." Also, the phrase *is equal to* becomes "$=$." We now have "3 times x = 10 more than x." The word *times* signals multiplication and the words *more than*, a backward phrase, signals addition. The problem becomes $3x = x + 10$. Now that we have an algebraic equation, we are ready to solve. Subtract x from both sides of the equation: $3x - x = x - x + 10$, $2x = 10$. Divide both sides of the equation by 2: $2x \div 2 = 10 \div 2$, $x = 5$.

CAUTION

ALGEBRAIC EQUATIONS CAN get a little trickier when keywords make it difficult to tell where to put certain numbers. Backward phrases like *more than* and *fewer than* don't just affect the order of numbers and variables—they affect which side of the equals sign numbers and variables should be placed.

Example

> Half of a number is twenty-one less than twice the number. What is the number?

Write *a number* as "x" and *twenty-one* as "21": Half of x is 21 less than twice x. If half of x is 21 less than twice x, then for these two quantities to be equal, we must either add 21 to the side of the equation that has half of *x*, or subtract 21 from the side of the equation that has twice *x*. We can do either, and the equation will still be correct. Which one should we do? Use the equals sign as a guide. Replace *is* with the equals sign: Half of *x* = 21 less than twice *x*. The backward phrase *less than* signals subtraction, so the right side of the equation becomes twice *x* – 21. *Twice x* means "*x* multiplied by 2," and *half of* x means "*x* divided by 2, or *x* multiplied by one-half." Our equation is $0.5x = 2x - 21$. Subtract $2x$ from both sides of the equation, and we are left with $-1.5x = -21$. Divide both sides of the equation by –1.5, and $x = 14$.

PRACTICE LAP

9. If the quotient of a number and nine is twelve, what is the number?

10. If the product of a number and ten is twenty more than fifty, what is the number?

11. Eight greater than a number is equal to three times the number. Find the number.

12. Six times a number plus three is the same as two fewer than seven times the number. What is the number?

13. One-third of a number equals four less than half the number. What is the number?

14. Five less than the square of a number is one more than five times the number. Find the two values that could be the number.

15. Three times the square of a number is fourteen more than the number. What two values could that number be?

ALGEBRAIC INEQUALITIES: SUBJECT REVIEW

We can write and solve algebraic inequalities just as we wrote and solved algebraic equations.

An **algebraic inequality** is a number sentence in which the algebraic expression on the left side is greater than, greater than or equal to, less than, or less than or equal to the algebraic expression on the right side. The number sentence $3x + 5 < 1$ is an algebraic inequality.

We solve an inequality just as we solve an equation, by isolating the variable on one side of the inequality using addition, subtraction, multiplication, division, or exponents. Since an inequality does not describe a single number, the solution to an inequality is a set of values rather than a single value. Note: There is, however, one key difference to solving an inequality. If you divide both sides of an inequality by a negative number, you must reverse the inequality symbol. If $-5x < 15$, then $x > -5$.

Inequalities contain the phrases *is less than, is less than or equal to, is greater than,* or *is greater than or equal to.* Be sure to spot the entire phrase in context. *Five is three less than eight* is "$5 = 8 - 3$," because the word *is* represents the equals sign and *less than* represents subtraction. But when those words are put together to form the phrase *is less than*, the phrase must be replaced with the less than symbol, $<$. In the same way, the phrase *seven is two greater than five* is "$7 = 5 + 2$," but *seven is greater than two plus four* is "$7 > 2 + 4$." Replace *is less than* with "$<$," *is less than or equal to* with "$\leq$," *is greater than* with "$>$," and *is greater than or equal to* with "$\geq$."

Example

> If seventeen more than five times a number is less than two, what set of values could be that number?

We begin this problem just as we would an algebraic equation. Write *seventeen* as "17," write *five* as "5," write *a number* as "x," and write *two* as "2": 17 more than five times x is less than 2. Replace *is less than* with "$<$." *Times* signals multiplication and *more than* is a backward phrase that signals addition. In this problem, *less than* does not signal subtraction, because it has been replaced with the $<$ symbol. The number sentence *17 more than five times x is less than 2* becomes "$5x + 17 < 2$." Subtract 17 from both sides of the inequality and divide by 5: $5x + 17 < 2$, $5x + 17 - 17 < 2 - 17$, $5x < -15$, $5x \div 5 < -15 \div 5$, $x < -3$. Since we divided by a positive number, we did not have to reverse the inequality symbol.

INSIDE TRACK

IF YOU SEE the word *negative* in an algebraic inequality word problem, be prepared to reverse the inequality symbol. You won't have to if you don't divide by a negative number, but it's easy to forget to reverse the inequality symbol, since this step is only required when working with certain inequalities. Circle the word *negative* to remind yourself to check to see if the inequality symbol must be reversed.

Example

Five plus negative six times a number is less than or equal to one less than negative four times that number. What set of values could the number be?

Write *five* as "5," write *negative six* as "−6," write *a number* as "x," write *one* as "1," and write *negative four* as "−4." The phrase *is less than or equal to* can be replaced with ≤": 5 plus −6 times $x \leq 1$ less than −4 times x. *Plus* signals addition and *times* signals multiplication. The words *less than* signal subtraction and not the less than sign, because the phrase is *less than*, not *is less, than.* We now have $5 + -6x \leq -4x - 1$, or $5 - 6x \leq -4x - 1$. Subtract 5 from both sides of the inequality and add $4x$ to both sides of the inequality. $-2x \leq -6$. Divide both sides by −2. Since we are dividing by a negative number, reverse the inequality symbol: $x \geq 3$.

INSIDE TRACK

IF YOU ARE confused about switching the direction of the inequality symbol, you can always move a negative x term to the other side of the inequality by adding to both sides of the inequality. For instance, if you have $-3x < 9$, you can add $3x$ to both sides of the inequality, giving you $0 < 3x + 9$. Subtract 9 from both sides, and you will have $-9 < 3x$. Now you can divide both sides of the inequality by 3, and you don't have to worry about changing the direction of the inequality symbol.

PACE YOURSELF

POUR SOME CEREAL into each of three bowls. Count the number of pieces of cereal in each bowl. Place two bowls on one side of the table and place the third bowl on the other side. What inequality symbol could you put in the center of the table to describe the relationship between the number of pieces of cereal on one side of the table versus the number of pieces of cereal on the other side of the table? What if you moved one of the bowls? Write a word problem to describe the cereal on the table. Explain how to solve the word problem if the amount of cereal in one bowl was unknown.

PRACTICE LAP

16. The difference between a number and two is greater than eight. Find the set of values that describes the number.

17. Ten times a number is greater than or equal to three times that number plus fourteen. What is the set of values that describes the number?

18. Nine less than a number multiplied by four is less than six times the number minus one. Find the set of values that describes the number.

19. Negative eight times a number is greater than or equal to negative five times the number plus eighteen. Describe the set of values that could be the number.

20. Three times a number, less than six, is less than the sum of twice the number and eleven. Find the set of values that describes the number.

SUMMARY

SOLVING ALGEBRA WORD problems can be a lot like solving multistep word problems. Algebraic equations and inequalities introduced us to a few new keywords and phrases, and we saw how backward phrases can appear in these types of problems, too. By converting words to numbers and symbols, we were able to turn word problems into plain algebra problems. Next, we'll get back to using the eight-step process as we learn how to solve word problems with fractions, decimals, and percents.

ANSWERS

1. Write *eight* as "8" and *a number* as "*x*": 8 more than *x*. *More than* is a phrase that signals addition, so 8 more than *x* is $x + 8$.

2. Write *a number* as "*x*" and *six* as "6": *x* divided by 6. *Divided by* is a phrase that signals division, so *x* divided by 6 is $\frac{x}{6}$.

3. Write *eleven* as "11," *a number* as "*x*," and *five* as "5": 11 times the sum of *x* and 5. We must find the sum of *x* and 5 first. *Sum* signals addition, so the sum of *x* and 5 is $x + 5$. *Times* signals multiplication, so 11 times *x* + 5 is $11(x + 5)$. We need parentheses so that 11 is multiplied by the sum and not just multiplied by *x*.

4. Write *a number* as "*x*," *nineteen* as "19," and *two* as "2": the quotient of *x* and the difference between 19 and 2. Before we can find the quotient, we must find the difference between 19 and 2. *Difference* signals subtraction, so the difference between 19 and 2 is $19 - 2$. *Quotient* signals division, so the quotient of a number and the difference between 19 and 2 is $\frac{x}{(19 - 2)}$. We need parentheses so that *x* is divided by the difference between 19 and 2 and not just divided by 19.

5. We can write one unknown number as *x* and the other unknown number as *y*. *Product* signals multiplication, so the product of two different numbers is *xy*. Twice that product is two multiplied by the product, or $2xy$.

6. Write *one* as "1," *six* as "6," and *a number* as "x." *Difference* signals subtraction, so the difference between 6 and x is $6 - x$. *More than* signals addition, so 1 more than $6 - x$ is $(6 - x) + 1$. The parentheses aren't required; either the subtraction or the addition can be performed first, but it is a good habit to group the operation that is to be performed first in parentheses.

7. Write *a number* as "x" and *five* as "5": the square of x, minus 5 times x. The square of x is x^2. *Times* signals multiplication, so 5 times x is $5x$. *Minus* signals subtraction, so x^2 minus $5x$ is $x^2 - 5x$.

8. Write *ten* as "10," *fifteen* as "15," and *a number* as "x": 10 fewer than 15 times the cube of x. The cube of x is x^3. *Times* signals multiplication, so 15 times x^3 is $15x^3$. The words *fewer than* signal subtraction and are also a backward phrase: 10 fewer than $15x^3$ is $15x^3 - 10$.

9. Write *a number* as "x," write *nine* as "9" and write *twelve* as "12." Replace the word *is* with the equals sign: the quotient of x and 9 = 12. *Quotient* signals division. The quotient of x and 9 is $\frac{x}{9}$. We now have $\frac{x}{9} = 12$. Solve the equation by multiplying both sides by 9: $x = 108$.

10. Write *a number* as "x," write *ten* as "10," write *twenty* as "20," and write *fifty* as "50." Replace the word *is* with the equals sign: the product of x and 10 = 20 more than 50. *Product* signals multiplication and *more than* signals addition. The product of x and 10 = 20 more than 50 becomes $10x = 50 + 20$. Solve the equation: $10x = 70$, $x = 7$.

11. Write *eight* as "8," write *a number* as "x," and write *three* as "3." Replace *is equal to* with the equals sign: 8 greater than x = 3 times x. *Greater than* signals addition, and *times* signals multiplication. The number sentence 8 greater than x = 3 times x becomes $x + 8 = 3x$. Subtract x from both sides of the equation: $x - x + 8 = 3x - x$, $8 = 2x$. Divide both sides of the equation by 2: $\frac{8}{2} = \frac{2x}{2}$, $x = 4$.

12. Write *six* as "6," write *a number* as "x," write *two* as "2," and write *seven* as "7." Replace *is the same as* with the equals sign: 6 times x plus 3 = 2 fewer than 7 times x. *Times* signals multiplication, *plus* signals addition, and *fewer than* signals subtraction. Remember, *fewer than* is a backward phrase: 6 times x plus 3 = 2 fewer than 7 times x becomes $6x + 3 = 7x - 2$. Subtract $6x$ from both sides of the equation and add 2 to both sides of the equation: $6x + 3 = 7x - 2$, $3 = x - 2$, $x = 5$.

13. Write *one-third* as "$\frac{1}{3}$," write *a number* as "*x*," write *four* as "4," and write *half* as "$\frac{1}{2}$." Replace *equals* with the equals sign: $\frac{1}{3}x = 4$ less than $\frac{1}{2}x$. *Less than* signals subtraction. Remember, *less than* is a backward phrase: $\frac{1}{3}x = 4$ less than $\frac{1}{2}x$ becomes $\frac{1}{3}x = \frac{1}{2}x - 4$. Convert both fractions to sixths: $\frac{2}{6}x = \frac{3}{6}x - 4$. Subtract $\frac{3}{6}x$ from both sides of the equation: $-\frac{1}{6}x = -4$. Multiply both sides by -6: $x = 24$.

14. Write *five* as "5," write *a number* as "*x*," write *one* as "1," and write *five* as "5." Replace *is* with the equals sign: 5 less than the square of $x = 1$ more than 5 times *x*. *The square* of *x* tells us that we must raise *x* to the second power. The square of *x* is x^2. *Less than* signals subtraction, *times* signals multiplication, and *more than* signals addition. Remember, *less than* and *more than* are both backward phrases. Our equation is now $x^2 - 5 = 5x + 1$. Subtract $5x$ and 1 from both sides of the equation: $x^2 - 5x - 6$. Factor the equation and set each factor equal to 0: $(x - 6)(x + 1)$, $x - 6 = 0$, $x = 6$, $x + 1 = 0$, $x = -1$.

15. Write *three* as "3," write *a number* as "*x*," and write *fourteen* as "14." Replace *is* with the equals sign: 3 times the square of $x = 14$ more than *x*. *The square* of *x* tells us that we must raise *x* to the second power. The square of *x* is x^2. *Times* signals multiplication and *more than* signals addition. Remember, *more than* is a backward phrase: 3 times the square of $x = 14$ more than *x* becomes $3x^2 = x + 14$. Subtract *x* and 14 from both sides of the equation: $3x^2 - x - 14$. Factor the equation and set each factor equal to 0: $(3x - 7)(x + 2)$, $3x - 7 = 0$, $3x = 7$, $x = \frac{7}{3}$, $x + 2 = 0$, $x = -2$.

16. Write *a number* as "*x*," write *two* as "2," and write *eight* as "8." Replace *is greater than* with the greater than symbol: the difference between *x* and 2 > 8. *Difference* signals subtraction, so we have $x - 2 > 8$, and $x > 10$.

17. Write *ten* as "10," write *a number* as "*x*," write *three* as "3," and write *fourteen* as "14." Replace *is greater than or equal to* with the greater than or equal to symbol: 10 times $x \geq 3$ times *x* plus 1. *Times* signals multiplication and *plus* signals addition, so we now have: $10x \geq 3x + 14$, $7x \geq 14$, $x \geq 2$.

18. Write *nine* as "9," write *a number* as "x," write *four* as "4," write *six* as "6," and write *one* as "1." Replace *is less than* with the less than symbol: 9 less than x multiplied by 4 < 6 times x minus 1. *Less than* signals subtraction, but it is a backward phrase, so 9 less than x is $x - 9$. *Multiplied* signals multiplication and *minus* signals subtraction, so we now have: $4x - 9 < 6x - 1$, $-2x < 8$. Divide by -2 and reverse the inequality symbol: $x > -4$.

19. Write *negative eight* as "-8," write *a number* as "x," write *negative five* as "-5," and write *eighteen* as "18." Replace *is greater than or equal* with the greater than or equal to symbol: -8 times $x \geq -5$ times x plus 18. *Times* signals multiplication and *plus* signals addition, so we now have: $-8x \geq -5x + 18$. Add $5x$ to both sides: $-3x \geq 18$. Divide by -3 and reverse the inequality symbol: $x \leq -6$.

20. Write *three* as "3," write *a number* as "x," write *six* as "6," write *twice* as 2, and write *eleven* as "11." Replace *is less than* with the less than symbol: 3 times x, less than 5 < the sum of $2x$ and 11. *Times* signals multiplication and *sum* signals addition. *Less than* is a backward phrase that signals subtraction, so we now have: $6 - 3x < 2x + 11$. Add $3x$ to both sides: $6 < 5x + 11$. Subtract 11 from both sides and divide by 5: $-5 < 5x$, $-1 < x$.

Fractions, Percents, and Decimals

The word problems we've looked at so far have involved mostly integers. In this chapter, we'll look at word problems that involve fractions, percents, and decimals. We'll use the same strategies we used to solve integer word problems.

FRACTIONS: SUBJECT REVIEW

To add two **like fractions**, add the numerators of the fractions and keep the denominator.

$$\frac{1}{8} + \frac{3}{8} = \frac{4}{8}$$

To add two **unlike fractions**, find common denominators and then add the numerators.

$$\frac{1}{4} + \frac{1}{6} = \frac{3}{12} + \frac{2}{12} = \frac{5}{12}$$

To subtract like fractions, subtract the numerator of the second fraction from the numerator of the first and keep the denominator.

$$\frac{3}{8} - \frac{1}{8} = \frac{2}{8}$$

To subtract unlike fractions, find common denominators, subtract the numerator of the second fraction from the numerator of the first, and keep the denominator.

$$\frac{3}{5} - \frac{3}{8} = \frac{24}{40} - \frac{15}{40} = \frac{9}{40}$$

To multiply two fractions, like or unlike, multiply the numerators and multiply the denominators.

$$\frac{2}{5} \times \frac{3}{7} = \frac{6}{35}$$

To divide two fractions, like or unlike, multiply the first fraction by the reciprocal of the second fraction.

$$\frac{5}{6} \div \frac{3}{4} = \frac{5}{6} \times \frac{4}{3} = \frac{20}{18}$$

To convert an **improper fraction** into a **mixed number**, divide the numerator of the fraction by the denominator. The whole number part of that division is the whole number part of the mixed number. The remainder, if any, becomes the numerator of the fraction, and the denominator remains the same.

$$\frac{20}{18} = \frac{12}{18}$$

To reduce or simplify a fraction, find the **greatest common factor** of the numerator and denominator and divide the numerator and denominator by that number.

$$\frac{12}{15} = \frac{\frac{12}{3}}{\frac{15}{3}} = \frac{4}{5}$$

FUEL FOR THOUGHT

LIKE FRACTIONS HAVE the same denominator; $\frac{5}{6}$ and $\frac{4}{6}$ are like fractions. **Unlike fractions** have different denominators; $\frac{5}{6}$ and $\frac{1}{9}$ are unlike fractions. A **proper fraction** a value between 0 and 1 or between −1 and 0. The number in the numerator is usually less than the number in the denominator of the fraction. An **improper fraction** has a value greater than or equal to 1, or less than or equal to −1. A **mixed number** contains both an integer and a fraction. The **greatest common factor** of two numbers is the largest integer that both numbers can be divided by with no remainder.

Now that we've reviewed how to work with fractions, let's look at some fraction word problems. We can use any of the strategies we've learned to answer these questions.

Example

DeDe pours 1 cup of cereal into a bowl. She adds $\frac{1}{2}$ cup of milk from a container that contains $2\frac{3}{4}$ cups of milk. How many cups of milk are left in the container?

Let's use the eight-step process to answer this question.

Read the entire word problem.
We are given the amount of milk in the container and the amount that DeDe pours into her cereal bowl.
Identify the question being asked.
We are looking for how much milk is left in the container.
Underline the keywords.
The keyword *left* signals subtraction.
Cross out extra information and translate words into numbers.
We do not need to know that there is 1 cup of cereal in the bowl. That information will not help us solve this problem, so cross it out.
List the possible operations.

To find how much milk is left in the container, we have to subtract the original amount in the container from the amount that DeDe poured out.

Write number sentences for each operation.

$2\frac{3}{4} - \frac{1}{2}$

Solve the number sentences and decide which answer is reasonable.

Convert one-half to fourths and subtract:

$2\frac{3}{4} - \frac{2}{4} = 2\frac{1}{4}$

Since DeDe used only half a cup of milk, this answer seems reasonable.

Check your work.

We solved this problem using subtraction, so we must use addition to check our work. Add the number of cups of milk DeDe put in her cereal to the new amount of milk in the container. The sum should equal the original volume of milk in the container: $2\frac{1}{4} + \frac{2}{4} = 2\frac{3}{4}$. Our answer is correct.

Example

Tatiana spends an hour and a half at the gym four days a week. How many hours does she spend at the gym each week?

Let's solve this problem by drawing a picture. Draw a full circle and a half circle for each of the four days Tatiana goes to the gym. The full circle represents a full hour, and the half circle represents half an hour:

There are four full circles and four half circles. Each pair of half circles makes a whole circle, so we have six whole circles. Tatiana spends six hours in the gym each week.

Example

Mark buys $3\frac{1}{8}$ pounds of turkey and $2\frac{1}{10}$ pounds of bologna. He uses $\frac{4}{5}$ pounds to make a sandwich. How many pounds of meat does he have left?

Read the entire word problem.

We are given the number of pounds of turkey and bologna that Mark buys and the number of pounds of meat he uses to make a sandwich.

Identify the question being asked.

We are looking for how much meat he has left.

Underline the keywords.

The keyword *add* signals addition, and the keyword *left* signals subtraction.

Cross out extra information and translate words into numbers.

There is no extra information in this word problem.

List the possible operations.

First, we need to find how much meat Mark bought by adding the weight of the turkey to the weight of the bologna. Then, we will subtract the weight of the meat used to make the sandwich from this total.

Write number sentences for each operation.

$3\frac{1}{8} + 2\frac{1}{10}$

Solve the number sentences and decide which answer is reasonable.

Convert eighths and tenths to fortieths and add:

$3\frac{1}{8} + 2\frac{1}{10} = 3\frac{5}{40} + 2\frac{4}{40} = 5\frac{9}{40}$

Write number sentences for each operation.

Now that we have the total weight of the meat Mark bought, we can subtract the weight of the meat used to make the sandwich to find how much meat is left:

$5\frac{9}{40} - \frac{3}{4}$

Solve the number sentences and decide which answer is reasonable.

Convert fourths to fortieths and subtract:

$5\frac{9}{40} - \frac{3}{4} = 5\frac{9}{40} - \frac{30}{40} = 4\frac{19}{40}$. Mark has $4\frac{19}{40}$ pounds of meat left.

Check your work.

We solved this problem using subtraction and addition, so we must use addition and subtraction to check our work. Add the number of pounds of meat Mark used for his sandwich to the number of pounds of meat he had left: $4\frac{19}{40} + \frac{30}{40} = 5\frac{9}{40}$. Subtract from that total the number of pounds of bologna Mark bought, $2\frac{1}{10}$, and we should be left with the number of pounds of turkey Mark bought, $3\frac{1}{8}$: $5\frac{9}{40} - 2\frac{1}{10}$ $= 5\frac{9}{40} - 2\frac{4}{40} = 3\frac{5}{40} = 3\frac{1}{8}$ pounds of meat.

PRACTICE LAP

1. When Zoe's birthday party is over, $\frac{5}{8}$ of the cake remains. If Morgan eats $\frac{1}{8}$ of the cake, how much of the cake will be left?

2. Sue has $\frac{3}{5}$ ounces of almond extract. If she divides it equally over four batches of fudge, how many ounces of almond extract are in each batch of fudge?

3. Yves uses $\frac{2}{3}$ yards of string to tie a bundle of magazines. If he has six bundles to tie, how many yards of string does he need?

4. Every student in Sayda's class must bring in $\frac{7}{8}$ square feet of fabric for an art project. If there are 24 students in the class, and her teacher brings in an additional $3\frac{4}{7}$ square feet of fabric, how many total square feet of fabric will be used for the project?

5. Patrick spends three-fourths of an hour studying for his math test and half an hour studying for his science test. How many total minutes did Patrick spend studying?

PACE YOURSELF

HOPE BUYS WATER by the case. One water company sells a case of 24 bottles, each filled with $\frac{5}{9}$ liters of water. A second company sells a case of 20 bottles, each filled with $\frac{5}{7}$ liters of water. Which case contains more water? How did you figure that out? If you knew the price of each case, what would you do to figure out which case was the better deal?

DECIMALS: SUBJECT REVIEW

We add and subtract decimals just as we add and subtract whole numbers; we just need to be sure to line up the decimal points of the addends (or minuend and subtrahend). Place trailing and leading zeros to keep your columns straight.

$$
\begin{array}{r}
454.991 \\
+\ 3.540 \\
\hline
458.531
\end{array}
\qquad
\begin{array}{r}
7.0003 \\
-\ 0.5400 \\
\hline
6.4603
\end{array}
$$

We multiply two decimals just as we multiply two whole numbers, but the number of places to the right of the decimal point of our answer is equal to the sum of the number of places to the right of the decimal point in our factors.

$$
\begin{array}{r}
5.670 \\
\times\quad 2.34 \\
\hline
22680 \\
17010 \\
+\ 11340 \\
\hline
13.26780
\end{array}
$$

To divide two decimals, shift the decimal point to the right until the divisor is an integer. Shift the decimal point of the dividend the same number of places to the right. Divide and carry the decimal point into the quotient.

$$
\frac{1{,}110.772}{21.32} = \frac{111{,}077.2}{2{,}132}
$$

$$
\begin{array}{r}
52.1 \\
2{,}132\overline{)111{,}077.2}
\end{array}
$$

To round a decimal to a place, look at the digit to the right of that place. If the digit is 5 or greater, increase the number in the place by one and change every digit to the right to zero. If the digit is 4 or less, leave the number in the place and change every digit to the right to zero.

54.6078 rounded to the thousandths place is 54.6080; 54.6078 rounded to the tenths place is 54.6000.

We've already seen some word problems that involve decimals. Almost every problem that deals with money involves two decimal places.

Example

If a raffle ticket costs $1.75, how many tickets can be bought with a $20 bill?

We could solve this problem using the eight-step process, but let's use a table instead. By multiplying the number of tickets by $1.75, we can see exactly when the total exceeds $20.

Number of Tickets	Price Per Ticket	Total
1	× $1.75	$1.75
2	× $1.75	$3.50
3	× $1.75	$5.25
4	× $1.75	$7.00
5	× $1.75	$8.75
6	× $1.75	$10.50
7	× $1.75	$12.25
8	× $1.75	$14.00
9	× $1.75	$15.75
10	× $1.75	$17.50
11	× $1.75	$19.25
12	× $1.75	$21.00

Eleven tickets can be purchased for less than $20, but 12 tickets would cost more than $20. The most tickets that can be purchased is 11.

Example

What is 4.7 less than 2.9 times the quotient of 5.75 and 0.25?

Break the problem down into single operations, placing parentheses around each. We are looking for 4.7 less than a number: ( ) – 4.7. That number is 2.9 times another number: $[2.9 \times (\ )] - 4.7$. The missing number is the quotient of 5.75 and 0.25, which is $\frac{5.75}{0.25}$: $[2.9 \times (\frac{5.75}{0.25})] - 4.7$. Start by finding the answer to the operation in the innermost parentheses: $\frac{5.75}{0.25} = 23$, so the problem becomes $(2.9 \times 23) - 4.7$. $2.9 \times 23 = 66.7$, so the problem is now $66.7 - 4.7 = 62$.

Example

The atomic mass of one atom of hydrogen is 1.00794 grams, the mass of one atom of sulfur is 32.065 grams, and the mass of one atom of oxygen is 15.994 grams. What is the weight of one molecule of sulfuric acid, which contains two hydrogen atoms, one sulfur atom, and four oxygen atoms?

Read the entire word problem.
We are given the masses of one atom of hydrogen, one atom of sulfur, and one atom of oxygen.
Identify the question being asked.
We are looking for the total weight of two hydrogen atoms, one sulfur atom, and four oxygen atoms.
Underline the keywords.
There are no keywords in this problem.
Cross out extra information and translate words into numbers.
There is no extra information in this problem.
List the possible operations.
We are given the weights of three different elements and the number of atoms we have of each element. To find the total weight, we must multiply the weight of one atom of each element by the number of atoms we have of that element. Since we have only one sulfur atom, we won't need to multiply that weight. We will need two multiplication sentences and one addition sentence.
Write number sentences for each operation.
First, find the weight of the hydrogen atoms. There are two, so multiply the weight of a hydrogen atom by 2:
1.00794×2
Solve the number sentences and decide which answer is reasonable.
$1.00794 \times 2 = 2.01588$ grams

Write number sentences for each operation.

Next, find the weight of the oxygen atoms. There are four, so multiply the weight of an oxygen atom by 4:

15.994 × 4

Solve the number sentences and decide which answer is reasonable.

15.994 × 4 = 63.976 grams

Write number sentences for each operation.

Finally, add the weights of the hydrogen, sulfur, and oxygen atoms:

2.01588 + 32.065 + 63.976

Solve the number sentences and decide which answer is reasonable.

2.01588 + 32.065 + 63.976 = 98.05688 grams

Check your work.

We twice used multiplication and addition to solve this problem, so we will twice use subtraction and division to check our work. Subtract the weight of the sulfur atom from the total weight: 98.05688 − 32.065 = 65.99188 grams, the weight of the hydrogen atoms. Now subtract 2.01588 grams, the combined weight of the two hydrogen atoms: 65.99188 − 2.01588 = 63.976 grams, the weight of the oxygen atoms. Divide the weight of the hydrogen atoms by 2 and divide the weight of the oxygen atoms by 2. We should be left with the atomic mass of each element: $\frac{2.01588}{2}$ = 1.00794 grams and $\frac{63.976}{4}$ = 15.994 grams.

INSIDE TRACK

DECIMALS CAN BE easier to work with than fractions, especially when you have to add or subtract unlike fractions. When a problem contains fractions that are easy to convert to decimals, such as halves, fourths, or tenths, turn the problem into a decimal problem—it may make your operations a lot easier!

6. The average rainfall per day for November in Sunnydale was 1.304 inches. How many total inches of rain fell in Sunnydale in November?

7. Find 3.27 more than the product of 9.3 less than 12.007 and the difference between 3.33 and 0.1.

8. A jar contains 0.346 liters of water. Every hour, 0.106 liters are added to the jar and 0.055 liters evaporate. How much water is in the jar after five hours?

9. Elle makes $43.75 per day at her part-time job. If she wants to buy a stereo that costs $600, how many full days must she work?

10. A multivitamin contains 0.015 grams of vitamin E. If a jar holds 36 vitamins, how many grams of vitamin E are in 7.5 jars of vitamins?

PERCENTS: SUBJECT REVIEW

We can write a decimal as a percent by moving the decimal point two places to the left and adding the percent sign.

0.23 = 23%

We can write a percent as a decimal by moving the decimal point two places to the right and removing the percent sign.

4% = 0.04

We can find a percent of a number by multiplying that number by that percent.

30% of 50 is 15 because 50 × 0.30 = 15.

We can find what percent one number is of a second number by dividing the first number by the second number.

2 is 40% of 5 because $\frac{2}{5} = 0.40 = 40\%$.

The percent increase or decrease from one number to another is the difference between the original number and the new number divided by the original number.

From 10 to 12 is a 20% increase because $\frac{(12-10)}{10} = \frac{2}{10} = 0.20 = 20\%$.

From 24 to 18 is a 25% decrease because $\frac{(24-18)}{24} = \frac{6}{24} = 0.25 = 25\%$.

If a number is increased by some percent, we can multiply that number by the percent to find by how much the number was increased.

If 30 is increased by 10%, it is increased by $30 \times 0.1 = 3$.

If 24 is decreased by 50%, it is decreased by $24 \times 0.5 = 12$.

Percent word problems almost always contain the word *percent*. When you see the word *percent*, use the context of the problem to decide if you must:

1. Find a percent of a number.
2. Find what percent one number is of a second number.
3. Find percent increase, either from one number to another, or a new value after a percent increase.
4. Find percent decrease, either from one number to another, or a new value after a percent decrease.

Example

Kevin must trim the grass of the football field, which is 48,000 square feet. If he cuts 35% of the grass before lunch, how much of the grass has he cut?

This is the first type of percent problem: finding the percent of a number. In this example, we must find 35% of 48,000. Convert 35% to a decimal and multiply by 48,000: 35% = 0.35; thus $0.35 \times 48,000 = 16,800$ square feet.

Percent word problems, like other word problems, can require more than one step.

Example

If Kevin cuts 30% of the grass before lunch and $\frac{2}{5}$ of the grass after lunch, how much grass does he have left to cut?

This is the same type of percent problem, but you must convert both 35% and $\frac{2}{5}$ to decimals, multiply each by 48,000, and subtract both products from 48,000 to find how much grass Kevin has left to cut. 30% = 0.3, 0.3 × 48,000 = 14,400. $\frac{2}{5}$ = 0.4, 0.4 × 48,000 = 19,200. 48,000 – 14,400 – 19,200 = 14,400 square feet.

If that last example seemed a little complicated, don't worry—we can use pictures and the eight-step process to solve these word problems, too. Let's look again at the last example.

The following diagram represents a football field, divided into ten equal sections:

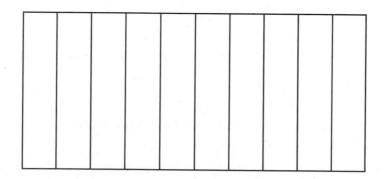

Why did we divide the field into ten sections? Kevin cuts 30%, which is $\frac{30}{100}$ or $\frac{3}{10}$ of the grass, before lunch, and he cuts $\frac{2}{5}$, which is the same as $\frac{4}{10}$, after lunch. Since both 30% and $\frac{2}{5}$ can be written as a fraction over ten, we can divide the field into ten sections, and shade the parts of the field that Kevin has cut. Shade three sections, since Kevin cut $\frac{3}{10}$ of the grass before lunch, and shade another four sections, since Kevin cut $\frac{4}{10}$ after lunch:

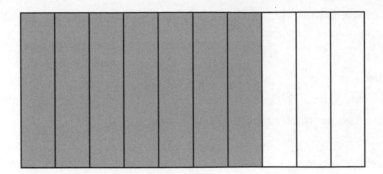

That leaves just three sections, or $\frac{3}{10}$, of the field left to cut. $\frac{3}{10}$ = 0.3, 0.3 × 48,000 = 14,400 square feet. Kevin has 14,400 square feet of grass left to cut.

Now, let's try a similar problem using the eight-step process.

Example

It takes 20 minutes for Frank to get to school every day. Today, it took Frank 15% longer to get to school because of traffic. How many minutes did it take for Frank to get to school?

Read the entire word problem.
We are given how long it usually takes for Frank to get to school, and the percentage increase of that amount of time.
Identify the question being asked.
We are looking for the number of minutes it took Frank to get to school today.
Underline the keywords.
There are no keywords in this problem, but the problem does contain a % symbol, so we will likely have to use a percent formula.
Cross out extra information and translate words into numbers.
There is no extra information in this problem.
List the possible operations.
Since we are told that it will take 15% longer to get to school, the number of minutes it takes Frank to get to school is increasing by 15%. We must use a percent increase formula.
Write number sentences for each operation.

Convert 15% to a decimal and multiply it by the number of minutes it usually takes Frank to get to school:

20 × 0.15

Solve the number sentences and decide which answer is reasonable.

20 × 0.15 = 3

Write number sentences for each operation.

It will take Frank three minutes longer to get to school today. To find out how long it took him to get to school, add three to the usual number of minutes it takes him to get to school:

20 + 3

Solve the number sentences and decide which answer is reasonable.

20 + 3 = 23 minutes

Check your work.

Since we found the increase in the number of minutes by multiplying by 0.15, divide the new number of minutes by 1.15, and we should have the original number of minutes: $\frac{23}{1.15}$ = 20 minutes.

INSIDE TRACK

IN THE LAST example, we found that 20 increased by 15% is equal to 23 by multiplying 20 by 0.15 and then adding the result to 20. We can find a 15% increase in 20 in one step if we multiply 20 by 1.15. Multiplying by (1 + 15%) means that our product will be 20 plus 15% of 20, which equals 23.

PRACTICE LAP

DIRECTIONS: Use the following information to answer questions 11–13:

Dan buys a music album that is 120 minutes long.

11. If Dan listens to 45% of the album, how many minutes of the album has he heard?

12. If Dan listens to 48 minutes of the album, what percent of the album has he heard?

13. If Dan has heard all but 15 minutes of the album, what percent of the album has he heard?

14. Aiden hauled 9,800 pounds of sand on Monday. If he hauls 8,700 pounds on Tuesday, how much less, by percent, did he haul on Tuesday? Round your answer to the nearest tenth of one percent.

15. Judi swims 25 laps every day. If she swims 12% more laps today, how many laps does she swim today?

16. Kellyann was 35 inches tall when she was three years old. If she is 65 inches tall now, by what percent has her height increased? Round your answer to the nearest tenth of one percent.

17. Jeffrey could bench-press 70 pounds in the seventh grade. A year later, he could press 20% more. As a ninth grader, he can press 25% more than he could as an eighth grader. How much can Jeffrey bench-press now?

PACE YOURSELF

A DECADE IS ten years. What percent of your age is one decade? When you are a decade older than you are now, by what percent will your age have increased? If you compare your age a decade ago to now, by what percent has your age increased?

SUMMARY

IN THIS CHAPTER, you learned how to recognize fraction, decimal, and percent word problems and how to use the strategies we learned in earlier chapters to solve these kinds of problems. We used formulas to solve percent word problems.

ANSWERS

1. $\frac{5}{8}$ of a cake is shown here:

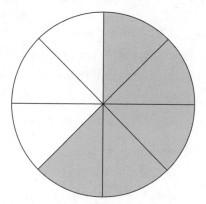

If we remove one of those eighths, $\frac{4}{8}$ are left:

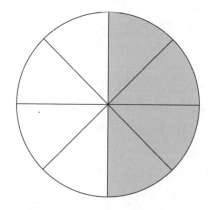

2. *Read the entire word problem.*

We are given the amount of almond extract that Sue has and the number of batches of fudge that she makes.

Identify the question being asked

We are looking for how much almond extract is in each batch.

Underline the keywords.

The keyword *divides* signals division.

Cross out extra information and translate words into numbers.

There is no extra information in this problem.

List the possible operations.

The keyword *each* also appears in this problem, but the word *divides* clearly tells us that we need to use division.

Write number sentences for each operation.

$$\frac{\left(\frac{3}{5}\right)}{4}$$

Solve the number sentences and decide which answer is reasonable.

The reciprocal of 4 is $\frac{1}{4}$, so multiply $\frac{3}{5}$ by $\frac{1}{4}$:

$$\frac{\left(\frac{3}{5}\right)}{4} = \frac{3}{5} \times \frac{1}{4} = \frac{3}{20} \text{ ounces}$$

Check your work.

We solved this problem using division, so we must use multiplication to check our work. Multiply the number of ounces of almond extract in each batch by the number of batches, and that product should give us the total amount of almond extract: $\frac{3}{20} \times 4 = \frac{12}{20} = \frac{3}{5}$ ounces.

3. *Read the entire word problem.*

We are given the amount of string Yves needs to tie one bundle of magazines and the number of bundles he has to tie.

Identify the question being asked.

We are looking for how many yards of string he needs.

Underline the keywords.

There are no keywords in this problem.

Cross out extra information and translate words into numbers.

There is no extra information in this problem.

List the possible operations.

We're given the number of yards of string needed to tie one bundle, so we will have to multiply that value by 6 to find the amount needed to tie six bundles.

Write number sentences for each operation.

$$\frac{2}{3} \times 6$$

Solve the number sentences and decide which answer is reasonable.

$$\frac{2}{3} \times 6 = \frac{12}{3} = 4 \text{ yards}$$

Check your work.

We solved this problem using multiplication, so we must use division to check our work. Divide the total number of yards by the number of yards needed to tie one bundle. This should give us the number of bundles Yves tied: $\frac{4}{\left(\frac{2}{3}\right)} = 4 \times \frac{3}{2} = 6$ bundles.

4. *Read the entire word problem.*

We are given the amount of fabric each student must bring and the amount of fabric the teacher is bringing.

Identify the question being asked.

We are looking for the total square feet of fabric.

Underline the keywords.

The keyword *every* can signal multiplication or division, and the keyword *additional* signals addition.

Cross out extra information and translate words into numbers.

There is no extra information in this problem.

List the possible operations.

If one student must bring $\frac{7}{8}$ square feet of fabric, then to find how many square feet 24 students must bring, we have to multiply $\frac{7}{8}$ by 24. Once we have that answer, we can add to it the amount of fabric the teacher brings.

Write number sentences for each operation.

$\frac{7}{8} \times 24$

Solve the number sentences and decide which answer is reasonable.

$\frac{7}{8} \times 24 = \frac{168}{8} = 21$ square feet

Write number sentences for each operation.

The students bring 21 square feet of fabric. Their teacher brings an additional $3\frac{4}{7}$ square feet, so we must add that to 21 to find the total:

$3\frac{4}{7} + 21$

Solve the number sentences and decide which answer is reasonable.

$3\frac{4}{7} + 21 = 24\frac{4}{7}$ square feet

Check your work.

We solved this problem using multiplication and addition, so we must use subtraction and division to check our work. Subtract the number of square feet of fabric brought by the teacher from the total. This should give us the amount of fabric brought by the students: $24\frac{4}{7} - 3\frac{4}{7}$ = 21 square feet. Divide 21 square feet by the number of children in the class. This should give us the number of square feet each student brought: $\frac{21}{24} = \frac{7}{8}$ square feet.

5. *Read the entire word problem.*

We are given the amount of time Patrick spent studying for his math test and for his science test.

Identify the question being asked.

We are looking for the total time, in minutes, spent studying.

Underline the keywords.

The keyword *total* signals addition.

Cross out extra information and translate words into numbers.

There is no extra information in this problem, but three-fourths of an hour and half an hour must be translated into fractions. Peter spent $\frac{3}{4}$ hours studying for his math test and $\frac{1}{2}$ hours studying for his science test. Also, we must express our answer in minutes, not hours. Remember, there are 60 minutes in an hour.

List the possible operations.

We must add the two fractions to find the total number of hours Patrick spent studying, and then multiply that sum by 60 to find the total number of minutes he spent studying.

Write number sentences for each operation.

$\frac{3}{4} + \frac{1}{2}$

Solve the number sentences and decide which answer is reasonable.

Convert halves to fourths and add:

$\frac{3}{4} + \frac{1}{2} = \frac{3}{4} + \frac{2}{4} = \frac{5}{4}$ hours

Write number sentences for each operation.

Multiply the number of hours by 60 to find the number of minutes Patrick spent studying:

$\frac{5}{4} \times 60$

Solve the number sentences and decide which answer is reasonable.

$\frac{5}{4} \times 60 = \frac{300}{4} = 75$ minutes

Check your work.

We solved this problem using addition and multiplication, so we must use division and subtraction to check our work. Divide the total number of minutes by 60 to convert the time spent studying from minutes to hours: $\frac{75}{60} = 1\frac{15}{60} = 1\frac{1}{4}$. Subtract the number of hours Patrick spent studying for his math test, and we should be left with the number of hours he spent studying for his science test: $1\frac{1}{4} - \frac{3}{4} = \frac{2}{4} = \frac{1}{2}$ hour.

6. *Read the entire word problem.*

We are given the average rainfall per day.

Identify the question being asked.

We are looking for the total rainfall for the month of November.

Underline the keywords.

The keywords *average* and *per* can signal multiplication or division.

Cross out extra information and translate words into numbers.

There is no extra information in this problem, but there are 30 days in the month of November. Replace the word *November* with "30 days."

List the possible operations.

Since we are given the amount of rainfall (on average) for one day and we are looking for the rainfall for 30 days, we must multiply.

Write number sentences for each operation.

1.304×30

Solve the number sentences and decide which answer is reasonable.

$1.304 \times 30 = 39.12$ inches

Check your work.

We solved this problem using multiplication, so we will use division to check our answer. Divide the total rainfall by 30, the number of days in November. This should equal the average rainfall per day: $\frac{39.12}{30} = 1.304$ inches.

7. We are looking for 3.27 more than a number, which is ( ) + 3.27. That number is the product of two numbers: ($\times$) + 3.27. The first of the two numbers is 9.3 less than 12.007, which is 12.007 – 9.3: [(12.007 – 9.3) $\times$] + 3.27. The other number is the difference between 3.33 and 0.1, which is 3.33 – 0.1. The number sentence is now [(12.007 – 9.3) $\times$ (3.33 – 0.1)] + 3.27. Do the subtraction in the innermost parentheses first: 12.007 – 9.3 = 2.707 and 3.33 – 0.1 = 3.23. The number sentence is now (2.707 $\times$ 3.23) + 3.27. Do the multiplication next: 2.707 $\times$ 3.23 = 8.74361. Finally, add 8.74361 and 3.27: 8.74361 + 3.27 = 12.01361.

8. We can use a table to solve this problem. Each hour, we add 0.106 and subtract 0.055, and then carry the new volume of water into the next row of the table:

Volume at start of hour	Volume added	Volume evaporated	Volume at end of hour
0.346	+ 0.106	– 0.055	0.397
0.397	+ 0.106	– 0.055	0.448
0.448	+ 0.106	– 0.055	0.499
0.499	+ 0.106	– 0.055	0.550
0.550	+ 0.106	– 0.055	0.601

After five hours, there are 0.601 liters in the jar.

9. We can use a table to solve this problem. Multiply the number of days worked by $43.75, increasing the number of days until the total earned reaches $600. We can tell that she must work more than just a few days, so we skip ahead to day five and then skip ahead to day ten. Once we see that the total is beginning to approach $600, we increase the number of days by one:

Number of days worked	Pay per day	Total earned
1	× $43.75	$43.75
5	× $43.75	$218.75
10	× $43.75	$437.50
11	× $43.75	$481.25
12	× $43.75	$525.00
13	× $43.75	$568.75
14	× $43.75	$612.50

After 13 days, Elle does not have enough for the stereo, so she must work 14 days to have enough money to buy the stereo.

10. *Read the entire word problem.*
We are given the number of grams of vitamin E in one vitamin, the number of vitamins in a jar, and the number of jars.
Identify the question being asked.
We are looking for the total number of grams of vitamin E.
Underline the keywords.
There are no keywords in this problem.
Cross out extra information and translate words into numbers.
There is no extra information in this problem.
List the possible operations.
Since we are given the number of grams in one vitamin, we must multiply to find the number of grams in 36 vitamins, or one jar of vitamins. We must multiply again to find the number of grams of vitamin E in 7.5 jars.
Write number sentences for each operation.
First, find the number of grams in one jar of 36 vitamins:
0.015×36
Solve the number sentences and decide which answer is reasonable.
$0.015 \times 36 = 0.54$ grams

Write number sentences for each operation.

Now find the number of grams in 7.5 jars of vitamins:

0.54 × 7.5

Solve the number sentences and decide which answer is reasonable.

0.54 × 7.5 = 4.05 grams

Check your work.

We solved this problem using multiplication twice, so we will use division twice to check our answer. Divide the total number of grams of vitamin E by 7.5 to find the number of grams in one jar: $\frac{4.05}{7.05}$ = 0.54 grams. Divide this number by 36, the number of vitamins in one jar, and this should give us the number of grams in one multivitamin: $\frac{0.54}{36}$ = 0.015 grams. Our answer is correct.

11. *Read the entire word problem.*

    We are given the length of the album in the directions, and we are given the percent of the album to which Dan has listened.

    Identify the question being asked.

    We are looking for the number of minutes of the album Dan has heard.

    Underline the keywords.

    There are no keywords in this problem, but the problem does contain a % symbol, so we will likely have to use a percent formula.

    Cross out extra information and translate words into numbers.

    There is no extra information in this problem.

    List the possible operations.

    Dan listens to 45% of a 120-minute album, so we must find 45% of 120.

    Write number sentences for each operation.

    Convert 45% to a decimal and multiply it by the length of the album:

    120 × 0.45

    Solve the number sentences and decide which answer is reasonable.

    120 × 0.45 = 54 minutes

    Check your work.

    We multiplied to find our answer, so we will divide to check our work. Since Dan listened to 54 minutes of the album, which is 45% of the album, divide 54 by 0.45 to find the full length of the album: $\frac{54}{0.45}$ = 120 minutes.

12. *Read the entire word problem.*

    We are given the length of the album in the directions, and we are given the number of minutes to which Dan has listened.

Identify the question being asked.

We are looking for the percent of the album Dan has heard.

Underline the keywords.

There are no keywords in this problem, but the problem does contain the word *percent*, so we will likely have to use a percent formula.

Cross out extra information and translate words into numbers.

There is no extra information in this problem.

List the possible operations.

Dan listens to 48 minutes of a 120-minute album, so we must find what percent 48 is of 120.

Write number sentences for each operation.

Divide 48 by 120 to find what percent 48 is of 120:

$\frac{48}{120}$

Solve the number sentences and decide which answer is reasonable.

$\frac{48}{120} = 0.4 = 40\%$

Check your work.

We divided to find our answer, so we will multiply to check our work. If 48 minutes is 40% of the album, then 120 multiplied by 40%, or 0.4, should equal 48: $120 \times 0.4 = 48$ minutes. Our answer is correct.

13. *Read the entire word problem.*

We are given the length of the album in the directions, and we are given the number of minutes to which Dan has not listened. We are looking for the percent of the album Dan has heard.

Underline the keywords.

There are no keywords in this problem, but the problem does contain the word *percent*, so we will likely have to use a percent formula.

Cross out extra information and translate words into numbers.

There is no extra information in this problem.

List the possible operations.

Dan has heard all but 15 minutes of the album, so we must first subtract 15 from the length of the album, and then find what percent that number is of the total number of minutes.

Write number sentences for each operation.

First, subtract 15 from 120 to find the number of minutes Dan has heard:

$120 - 15$

Solve the number sentences and decide which answer is reasonable.

120 − 15 = 105

Write number sentences for each operation.

Divide 105 by the total length of the album, 120, to find the percent of the album that Dan has heard:

$\frac{105}{120}$

Solve the number sentences and decide which answer is reasonable.

$\frac{105}{120} = 0.875 = 87.5\%$

Check your work.

We subtracted and divided to find our answer, so we will multiply and add to check our work. If 105 minutes is 87.5% of the album, then 120 multiplied by 87.5%, or 0.875, should equal 105: 120 × 0.875 = 105 minutes. The number of minutes Dan has heard, 105 minutes, plus the number of minutes he has not heard, 15, should equal the total length of the album: 105 + 15 = 120 minutes, the total length of the album.

14. *Read the entire word problem.*

We are given the number of pounds Aiden hauls on Monday and on Tuesday.

Identify the question being asked.

We are looking for the percent decrease in the amount he hauled from Monday to Tuesday.

Underline the keywords.

The keyword *less* often signals subtraction.

Cross out extra information and translate words into numbers.

There is no extra information in this problem.

List the possible operations.

This problem asks how much less Aiden hauls on Tuesday than Monday by percent, which means that we must find the percent decrease from 9,800 to 8,700. The percent decrease is found by taking the difference between the original number and the new number and dividing by the original number.

Write number sentences for each operation.

The original number is 9,800 pounds, and the new number is 8,700. Plug these values into the formula:

$\frac{(9,800 - 8,700)}{9,800}$

Solve the number sentences and decide which answer is reasonable.

$\frac{(9,800 - 8,700)}{9,800} = \frac{1,100}{9,800}$, which ≈ 0.1122, or 11.2% rounded to the nearest tenth of one percent.

Check your work.

If 9,800 to 8,700 $\approx$ an 11.2% decrease, then increasing 8,700 by 11.2% should give us approximately 9,800. Instead of multiplying 8,700 by 0.112 and adding 8,700, we will simply multiply 8,700 by 1.112: 8,700 $\times$ 1.112 = 9,674.4, which $\approx$ 9,800.

15. *Read the entire word problem.*

We are given the number of laps Judi swims every day and the percentage increase in the number of laps she swam today.

Identify the question being asked.

We are looking for the number of laps she swam today.

Underline the keywords.

There are no keywords in this problem, but the problem does contain a % symbol, so we will likely have to use a percent formula.

Cross out extra information and translate words into numbers.

There is no extra information in this problem.

List the possible operations.

If Judi swam 12% more laps today, then the number of laps she swam increased by 12%. We will use a percent increase formula. Once we find how many more laps she swam today, we can add it to 25 to find the number of laps she swam.

Write number sentences for each operation.

If a number is increased by some percent, we can multiply that number by the percent to find by how much the number was increased:

25 $\times$ 0.12

Solve the number sentences and decide which answer is reasonable.

25 $\times$ 0.12 = 3 laps. Judi swam three more laps today.

Write number sentences for each operation.

Add the number of additional laps she swam to the usual number of laps she swims:

3 + 25

Solve the number sentences and decide which answer is reasonable.

3 + 25 = 28 laps

Check your work.

Since 28 is a 12% increase, divide 28 by 1.12 to find the number of laps Judi usually swims: $\frac{28}{1.12}$ = 25 laps.

16. *Read the entire word problem.*

We are given Kellyann's height when she was three years old and her height now.

Identify the question being asked.

We are looking for the percent by which her height has increased.

Underline the keywords.

There are no keywords in this problem, but the problem does contain the word *percent*, so we will likely have to use a percent formula.

Cross out extra information and translate words into numbers.

We are given the age three years old, but we don't need the number 3 to solve this problem. Kellyann's age is given there to show how her height has increased.

List the possible operations.

The problem asks us to find by what percent her height has increased. The percent increase is found by taking the difference between the original number and the new number and dividing by the original number.

Write number sentences for each operation.

The original number is 35 inches, and the new number is 65 inches. Plug these values into the formula:

$$\frac{(65 - 35)}{35}$$

Solve the number sentences and decide which answer is reasonable.

$\frac{(65 - 35)}{35} = \frac{30}{35}$, which ≈ 0.8571, or 85.7% rounded to the nearest tenth of one percent.

Check your work.

If 35 to 65 ≈ an 85.7% increase, then multiplying 35 by 185.7%, or 1.857, should give us approximately 65: 35 × 1.857 = 64.995, which ≈ 65.

17. *Read the entire word problem.*

We are given how much Jeffrey could bench-press in seventh grade, and his percent increases in that number from seventh grade to eighth grade and from eighth grade to ninth grade.

Identify the question being asked.

We are looking for how much Jeffrey can bench-press now.

Underline the keywords.

There are no keywords in this problem, but the problem does contain the % symbol, so we will likely have to use a percent formula.

Cross out extra information and translate words into numbers.

There is no extra information in this problem.

List the possible operations.

We need to find how much Jeffrey can bench-press as a ninth grader. We are given how much he can bench-press in seventh grade and the percentage increase in that weight to the amount he can bench-press in eighth grade. We must find how much he can bench-press in eighth grade, and then use that to find how much he can bench-press in ninth grade.

Write number sentences for each operation.

Convert 20% to a decimal. Before multiplying, add 1 to that decimal, so that after we multiply it by the number of pounds Jeffrey could bench-press in seventh grade, we won't need to add in order to find how many pounds Jeffrey could bench-press in eighth grade:

70×1.20

Solve the number sentences and decide which answer is reasonable.

$70 \times 1.20 = 84$ pounds

Write number sentences for each operation.

Jeffrey can bench-press 25% more as a ninth grader. Convert 20% to a decimal, and again, add 1 before multiplying. This will give us the weight Jeffrey can bench-press as a ninth grader:

84×1.25

Solve the number sentences and decide which answer is reasonable.

$84 \times 1.25 = 105$ pounds

Check your work.

Since we multiplied twice to find our answer, we will divide twice to check our work. Divide the weight Jeffrey could bench-press in ninth grade by 1 plus the percentage increase from eighth grade. This should give us how much Jeffrey could bench-press in eighth grade: $\frac{105}{1.25} = 84$ pounds. Divide this weight by 1 plus the percentage increase from seventh grade. This should give us how much Jeffrey could bench-press in seventh grade: $\frac{84}{1.2} = 70$ pounds. Our answer is correct.

Ratios and Proportions

We can use relationships between numbers or pairs of numbers to solve problems. For instance, let's say a deck of cards contains twice as many red cards as black cards. If we know the number of red cards, we can find the number of black cards, because we know the relationship between the number of red cards and the number of black cards. This kind of relationship is called a ratio. A **ratio** is a comparison, or relationship, between two numbers. Ratios can be shown using a colon or as a fraction. The ratio 3 to 2 can be written as 3:2 or $\frac{3}{2}$.

RATIOS: SUBJECT REVIEW

If there are 16 black pens and 12 blue pens on a desk, then the ratio of black pens to blue pens is 16:12. Ratio can be reduced just like fractions, since ratios are fractions. The greatest common factor of 16 and 12 is 4, so both numbers can be divided by 4. The ratio 16:12 is the same as the ratio 4:3.

THE ORDER OF the numbers in a ratio is important. The ratio 3:4 is not the same as the ratio 4:3. Remember, ratios are fractions, and the fraction $\frac{3}{4}$ is not equal to the fraction $\frac{4}{3}$. If you are describing the ratio of apples to oranges, place the number of apples first in your ratio.

If a word problem simply asks you to find or write a ratio, you may not need the eight-step process to find the answer. However, as with any kind of word problem, the eight-step process will work, and you should use it if you find that you cannot solve a word problem quickly.

Example

If there are 20 red cars and 24 blue cars in a parking lot, what is the ratio of red cars to blue cars?

This problem asks us to find a ratio, and it gives us the number of red cars and the number of blue cars. No computation is needed. The ratio of red cars to blue cars is 20:24. We can reduce the ratio to 5:6.

Example

A parking lot contains only red cards and blue cars. If there are 36 total cars and 16 of them are red, what is the ratio of blue cars to red cars?

Again, we are asked to write a ratio, but this time, we are not given the number of blue cars. We can find the number of blue cars by subtracting the number of red cars from the total number of cars: 36 − 16 = 20. We must be sure to write our ratio in the correct order. The problem asks us to find the ratio of blue cars to red cars, which is 20:16, or 5:4. Determining the question being asked can make the difference between a right or wrong answer.

PRACTICE LAP

DIRECTIONS: Answer the following questions and reduce all ratios.

1. If there are five adults and 25 students on a field trip, what is the ratio of adults to students?

2. If there are four roses and 12 daffodils in a vase, what is the ratio of daffodils to roses?

3. Every tree in Woodmere Park is either an oak or a maple. If there are 28 oak trees and 44 total trees, what is the ratio of oak trees to maple trees?

4. Ellie's purse contains 15 nickels and ten dimes. What is the ratio of nickels to total coins?

5. A box contains eight jelly doughnuts and four glazed doughnuts. What is the ratio of glazed doughnuts to total doughnuts?

PACE YOURSELF

FIND THE RATIO of $1 bills to $5 bills in your wallet or purse. What would that ratio be if you added three $5 bills to your wallet or purse?

PROPORTIONS: SUBJECT REVIEW

We can use ratios to solve problems by setting up a proportion. A **proportion** is a relationship between two equivalent ratios. A proportion usually contains a reduced ratio that represents a relationship between two quantities, and a larger, equivalent ratio that represents the exact values of those two quantities. The ratio 16:12 is equivalent to the ratio 4:3. These equivalent ratios can be expressed as a proportion: $\frac{16}{12} = \frac{4}{3}$.

If the ratio of quantity A to quantity B is 7:2, and the exact value of quantity A is 35, we can set up a proportion to solve for x, the exact value of quantity B. As a fraction, 7:2 is $\frac{7}{2}$. Set that fraction equal to the exact value of quantity A over the exact value of quantity B: $\frac{7}{2} = \frac{35}{x}$. Cross multiply and set the products equal to each other: $7x = 70$, $x = 10$. If the exact value of quantity A is 35, then exact value of quantity B is 10.

If the ratio of quantity A to quantity B is 7:2, and the exact total of quantity A and quantity B is 27, we can find the exact value of quantity A and quantity B by writing ratios that represent the part-to-whole relationship. If the ratio of quantity A to quantity B is 7:2, then 7 of every $7 + 2 = 9$ items are of quantity A. The ratio of quantity A to the whole is 7:9. The whole is 27, and exact number of quantity A is x: $\frac{7}{9} = \frac{x}{27}$. Cross multiply and set the products equal to each other: $9x = 189$, $x = 21$. If the total of quantity A and quantity B is 27, then the total of quantity A is 21. The total of quantity B can be found with the proportion $\frac{2}{9} = \frac{x}{27}$. $9x = 54$, $x = 6$. If the total of quantity A and quantity B is 27, then the total of quantity B is 6.

Now that we remember how to use ratios and proportions, let's look at how to recognize and solve ratio and proportion word problems.

Ratio word problems often contain the word *ratio* or *every*. You may also see the ratio expressed in colon form, such as 2:3. Once you've recognized a word problem as a ratio and proportion problem, decide if it is a part-to-part problem or a part-to-whole problem. A part-to-part problem gives you a ratio and the value of one quantity, and asks you for the value of the other quantity. A part-to-whole problem gives you a ratio and the total value of both quantities, and asks you for the value of one quantity.

INSIDE TRACK

ALTHOUGH THE KEYWORD *total* can often signal addition, in a ratio problem, the word *total* can signal a part-to-whole problem rather than a part-to-part problem.

Example

The ratio of peanuts to cashews in a jar of nuts is 5:6, and the exact number of peanuts in the jar is 25. How many cashews are in the jar?

Read the entire word problem.

We are given the ratio of peanuts to cashews and the exact number of peanuts.

Identify the question being asked.

We are looking for the number of cashews.

Underline the keywords and words that indicate formulas.

The word *ratio* indicates that we will either be finding a ratio or using a ratio to set up a proportion.

Cross out extra information and translate words into numbers.

There is no extra information in this problem.

List the possible operations.

This problem contains two quantities, of peanuts and of cashews. We are given the value of one quantity, and we are asked for the value of the other quantity. This is a part-to-part problem. We must set up a proportion that compares the ratio of peanuts to cashews to the actual numbers of peanuts and cashews.

Write number sentences for each operation.

The ratio of peanuts to cashews is 5:6, or $\frac{5}{6}$. Use x to represent the actual number of cashews, since that number is unknown. The ratio of actual peanuts to cashews is 25:x, or $\frac{25}{x}$. Set these fractions equal to each other:

$$\frac{5}{6} = \frac{25}{x}$$

Solve the number sentences and decide which answer is reasonable.

Cross multiply and solve for x:

$(6)(25) = 5x$, $150 = 5x$, $x = 30$. If there are 25 peanuts in the jar, then there must be 30 cashews in the jar.

Check your work.

The ratio of peanuts to cashews is 5:6, so we can check our answer by comparing that ratio to the actual numbers of peanuts and cashews. The actual numbers of peanuts and cashews should reduce to the ratio 5:6. The greatest common factor of 25 and 30 is 5, and 25:30 does reduce to 5:6. Our answer is correct.

BEFORE SETTING UP a proportion, be sure any given ratios are in reduced form. This will make cross multiplying and solving the proportion easier.

Example

Dermot finds that the ratio of basketball players to football players at his school is 3:15. If there are 90 total players, how many of the players are basketball players?

Read the entire word problem.
We are given the ratio of basketball players to football players and the total number of players.
Identify the question being asked.
We are looking for the number of basketball players.
Underline the keywords and words that indicate formulas.
The word *ratio* indicates that we will either be finding a ratio or using a ratio to set up a proportion. The word *total* is a signal that this is a part-to-whole ratio problem.
Cross out extra information and translate words into numbers.
There is no extra information in this problem.
List the possible operations.
This problem contains two quantities of basketball players and of football players. We are given the total of the quantities and we are asked for the value of one quantity. This is a part-to-whole problem. Since we are looking for the number of basketball players, we must set up a proportion that compares the ratio of basketball players to total players to the actual numbers of basketball players and total players.
Write number sentences for each operation.
The ratio of basketball players to football players is 3:15, which means that the ratio of basketball players to total players is 3:(15 + 3), which is 3:18. Reduce this ratio by dividing both numbers by 3: 3:18 = 1:6, or $\frac{1}{6}$. Use x to represent the actual number of basketball

players, since that number is unknown. There are 90 total players, so the ratio of actual basketball players to total players is x:90, or $\frac{x}{90}$. Set these fractions equal to each other:

$$\frac{1}{6} = \frac{x}{90}$$

Solve the number sentences and decide which answer is reasonable.
Cross multiply and solve for x:

$6x = 90$, $x = 15$. If there are 90 total players, then there must be 15 basketball players.

Check your work.

The ratio of basketball players to total players is 1:6. Check that the actual number of players and the total number of players are in the ratio 1:6; 15:90 = 1:6, so our answer is correct.

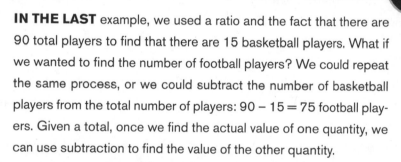

INSIDE TRACK

IN THE LAST example, we used a ratio and the fact that there are 90 total players to find that there are 15 basketball players. What if we wanted to find the number of football players? We could repeat the same process, or we could subtract the number of basketball players from the total number of players: 90 − 15 = 75 football players. Given a total, once we find the actual value of one quantity, we can use subtraction to find the value of the other quantity.

We've used proportions and the value of one quantity to find the value of another quantity, and we've used proportions and the total of two quantities to find the value of one quantity. We can also use proportions and the value of one quantity to find the total of two quantities.

Example

Caroline's summer camp has indoor days and outdoor days, depending on the weather. The ratio of indoor days to outdoor days was 3:17. If the camp had 34 outdoor days, for how many total days did Caroline attend camp?

Read the entire word problem.

We are given the ratio of indoor days to outdoor days and the actual number of outdoor days.

Identify the question being asked.

We are looking for the total number of days that Caroline attended camp.

Underline the keywords and words that indicate formulas.

The word *ratio* indicates that we will either be finding a ratio or using a ratio to set up a proportion. The word *total* is a signal that this is a part-to-whole ratio problem.

Cross out extra information and translate words into numbers.

There is no extra information in this problem.

List the possible operations.

Again, this problem contains two quantities, indoor days and outdoor days. We are given the value of one quantity, outdoor days, and we are asked for the total value. This is a part-to-whole problem. Since we are looking for the total number of days, we must set up a proportion that compares the ratio of outdoor days to total days to the actual numbers of outdoor days and total days.

Write number sentences for each operation.

The ratio of indoor days to outdoor days is 3:17, which means that the ratio of outdoor days to total days is 17:(17 + 3), which is 17:20, or $\frac{17}{20}$. Use x to represent the total number of days, since that number is unknown. There are 34 outdoor days, so the ratio of outdoor days to total days is 34:x, or $\frac{34}{x}$. Set these fractions equal to each other:

$$\frac{17}{20} = \frac{34}{x}$$

Solve the number sentences and decide which answer is reasonable.

Cross multiply and solve for x:

$17x = 680$, $x = 40$. If there are 34 outdoor days, then there must be 40 total days.

Check your work.

The ratio of outdoor days to total days is 17:20. Check that the actual number of outdoor days and the total number of days are in the ratio 17:20; 34:40 = 17:20, so our answer is correct.

6. The ratio of chairs to tables in the school cafeteria is 8:1. If there are 96 chairs in the cafeteria, how many tables are in the cafeteria?

7. The ratio of orange jelly beans to lemon jelly beans in a jar is 4:9. If there are 99 lemon jelly beans in the jar, how many orange jelly beans are in the jar?

8. There are 56 volunteers at the community center. If the ratio of paid employees to volunteers is 3:7, how many paid employees are at the community center?

9. Nick collects books. The ratio of his fiction books to his nonfiction books is 11:6. If Nick has 154 fiction books, how many nonfiction books does he have?

10. A card store sold 165 birthday cards and Valentine's Day cards in total on Sunday. If the ratio of birthday cards to Valentine's Day cards is 2:9, how many birthday cards did the store sell?

11. A rain forest contains many animals, including orangutans and gorillas. The ratio of orangutans to gorillas is 5:3. If the total number of orangutans and gorillas is 312, how many gorillas are in the rain forest?

12. A total of 88 students write for either the school newspaper or the yearbook. If the ratio of yearbook writers to newspaper writers is 13:9, how many students write for the yearbook?

13. A train conductor sells one-way tickets and round-trip tickets. The ratio of one-way tickets sold to round-trip tickets sold is 8:15. If the conductor sold 75 round-trip tickets, how many total tickets did he sell?

14. A magazine company hires part-time employees and full-time employees. The ratio of part-time employees to full-time employees is 5:14, and 168 of them are full-time employees. How many total people are employed at the magazine?

15. The ratio of Scottsdale residents to Wharton residents is 7:12. If 8,435 people live in Scottsdale, how many people live in Scottsdale and Wharton combined?

SUMMARY

THE WORD *RATIO* or the symbol ":" usually indicates that we will need a proportion to solve a word problem. Once we know that we are working on a ratio word problem, the word *total* can help us determine if it is a part-to-part problem or a part-to-whole problem.

ANSWERS

1. If there are five adults and 25 students, then the ratio of adults to students is 5 to 25, or 5:25. We can reduce this ratio by dividing both numbers by 5; 5:25 = 1:5.

2. If there are four roses and 12 daffodils, then the ratio of roses to daffodils is 4:12 and the ratio of daffodils to roses is 12:4. We can reduce this ratio by dividing both numbers by 3; 12:4 = 3:1.

3. If there are 28 oak trees and 44 total trees, and every tree is an oak or a maple, then there are 44 − 28 = 16 maple trees. The ratio of oak trees to maple trees is 28:16. We can reduce this ratio by dividing both numbers by 4; 28:16 = 7:4.

4. Ellie's purse contains 15 + 10 = 25 total coins. Since 15 of them are nickels, the ratio of nickels to total coins is 15:25. We can reduce this ratio by dividing both numbers by 5; 15:25 = 3:5.

5. The box contains 8 + 4 = 12 total doughnuts. Since four of them are glazed, the ratio of glazed doughnuts to total doughnuts is 4:12. We can reduce this ratio by dividing both numbers by 4; 4:12 = 1:3.

6. *Read the entire word problem.*

We are given the ratio of chairs to tables and the exact number of chairs.

Identify the question being asked.

We are looking for the number of tables.

Underline the keywords and words that indicate formulas.

The word *ratio* means that we will likely be using a ratio to set up a proportion.

Cross out extra information and translate words into numbers.

There is no extra information in this problem.

List the possible operations.

This problem contains two quantities, chairs and tables. We are given the value of one quantity, chairs, and we are asked for the value of the other quantity, tables. This is a part-to-part problem. We must set up a proportion that compares the ratio of chairs to tables to the ratio of the actual numbers of chairs to tables.

Write number sentences for each operation.

The ratio of chairs to tables is 8:1, or $\frac{8}{1}$. Use x to represent the actual number of tables, since that number is unknown. The ratio of actual chairs to tables is 96:x, or $\frac{96}{x}$. Set these fractions equal to each other:
$$\frac{8}{1} = \frac{96}{x}$$
Solve the number sentences and decide which answer is reasonable.

Cross multiply and solve for x:

$(96)(1) = 8x, 96 = 8x, x = 12$. If there are 96 chairs in the cafeteria, then there must be 12 tables in the cafeteria.

Check your work.

The ratio of chairs to tables is 8:1, so we can check our answer by comparing that ratio to the actual numbers of chairs to tables. The actual numbers of chairs and tables should reduce to the ratio 8:1. The greatest common factor of 96 and 12 is 8, and 96:12 does reduce to 8:1.

7. *Read the entire word problem.*

We are given the ratio of orange jelly beans to lemon jelly beans and the exact number of lemon jelly beans.

Identify the question being asked.

We are looking for the number of orange jelly beans.

Underline the keywords and words that indicate formulas.

The word *ratio* means that we will likely be using a ratio to set up a proportion.

Cross out extra information and translate words into numbers.

There is no extra information in this problem.

List the possible operations.

This problem contains two quantities, orange jelly beans and lemon jelly beans. We are given the value of one quantity, lemon jelly beans, and we are asked for the value of the other quantity, orange jelly beans. This is a part-to-part problem. We must set up a proportion that compares the ratio of orange jelly beans to lemon jelly beans to the ratio of the actual numbers of orange jelly beans to lemon jelly beans.

Write number sentences for each operation.

The ratio of orange jelly beans to lemon jelly beans is 4:9, or $\frac{4}{9}$. Use x to represent the actual number of orange jelly beans, since that number is unknown. The ratio of actual orange jelly beans to lemon jelly beans is x:99, or $\frac{x}{99}$. Set these fractions equal to each other:

$$\frac{4}{9} = \frac{x}{99}$$

Solve the number sentences and decide which answer is reasonable.

Cross multiply and solve for x:

$9x = (4)(44)$, $9x = 396$, $x = 44$. If there are 99 lemon jelly beans in the jar, then there must be 44 orange jelly beans in the jar.

Check your work.

The ratio of orange jelly beans to lemon jelly beans is 4:9, so the ratio of actual orange jelly beans to actual lemon jelly beans should reduce to 4:9. The greatest common factor of 44 and 99 is 11, and 44:99 reduces to 4:9.

8. *Read the entire word problem.*

We are given the ratio of paid employees to volunteers and the exact number of volunteers.

Identify the question being asked.

We are looking for the number of paid employees.

Underline the keywords and words that indicate formulas.

The word *ratio* means that we will likely be using a ratio to set up a proportion.

Cross out extra information and translate words into numbers.

There is no extra information in this problem.

List the possible operations.

This problem contains two quantities, paid employees and volunteers. We are given the value of one quantity, volunteers, and we are asked for the value of the other quantity, paid employees. This is a part-to-part problem. We must set up a proportion that compares the ratio of paid employees to volunteers to the ratio of actual paid employees to actual volunteers.

Write number sentences for each operation.

The ratio of paid employees to volunteers is 3:7, or $\frac{3}{7}$. Use x to represent the actual number of paid employees, since that number is unknown. The ratio of actual paid employees to actual volunteers is x:56, or $\frac{x}{56}$. Set these fractions equal to each other:

$\frac{3}{7} = \frac{x}{56}$.

Solve the number sentences and decide which answer is reasonable.

Cross multiply and solve for x:

$7x = (3)(56)$, $7x = 168$, $x = 24$. If there are 56 volunteers at the community center, then there must be 24 paid employees at the center.

Check your work.

The ratio of paid employees to volunteers is 3:7, so the ratio of actual paid employees to actual volunteers should reduce to 3:7. The greatest common factor of 24 and 56 is 8, and 24:56 reduces to 3:7.

9. *Read the entire word problem.*

We are given the ratio of fiction books to nonfiction books and the exact number of fiction books.

Identify the question being asked.

We are looking for the number of nonfiction books.

Underline the keywords and words that indicate formulas.

The word *ratio* means that we will likely be using a ratio to set up a proportion.

Cross out extra information and translate words into numbers.

There is no extra information in this problem.

List the possible operations.

This problem contains two quantities, fiction books and nonfiction books. We are given the value of one quantity, fiction books, and we are asked for the value of the other quantity, nonfiction books. This is a part-to-part problem. We must set up a proportion that compares the ratio of fiction books to nonfiction books to the ratio of actual fiction books to actual nonfiction books.

Write number sentences for each operation.

The ratio of fiction books to nonfiction books is 11:6, or $\frac{11}{6}$. Use x to represent the actual number of nonfiction books, since that number is unknown. The ratio of actual fiction books to actual nonfiction books is 154:x, or $\frac{154}{x}$. Set these fractions equal to each other:

$$\frac{11}{6} = \frac{154}{x}$$

Solve the number sentences and decide which answer is reasonable.

Cross multiply and solve for x:

$11x = (154)(6)$, $11x = 924$, $x = 84$. If Nick has 154 fiction books, then he has 84 nonfiction books.

Check your work.

The ratio of fiction books to nonfiction books is 11:6, so the ratio of actual fiction books to actual nonfiction books should reduce to 11:6. The greatest common factor of 154 and 84 is 14, and 154:84 reduces to 11:6.

10. *Read the entire word problem.*

We are given the ratio of birthday cards and Valentine's Day cards and the total number of cards sold.

Identify the question being asked.

We are looking for the number of birthday cards sold.

Underline the keywords and words that indicate formulas.

The word *ratio* means that we will likely be using a ratio to set up a proportion. The word *total* is a signal that this is a part-to-whole ratio problem.

Cross out extra information and translate words into numbers.

There is no extra information in this problem.

List the possible operations.

This problem contains two quantities, birthday cards and Valentine's Day cards. We are given the total of the two quantities, and we are looking for the value of one quantity. This is a part-to-whole problem. Since we are looking for the number of birthday cards sold, we must set up a proportion that compares the ratio of birthday cards sold to total cards sold to the ratio of actual birthday cards sold to actual total cards sold.

Write number sentences for each operation.

The ratio of birthday cards to Valentine's Day cards is 2:9, which means that the ratio of birthday cards to total cards is 2:(2 + 9), which is 2:11, or $\frac{2}{11}$. Use x to represent the number of birthday cards sold, since that number is unknown. There are 165 total cards sold, so the ratio of birthday cards sold to total cards sold is x:165, or $\frac{x}{165}$. Set these fractions equal to each other:

$$\frac{2}{11} = \frac{x}{165}$$

Solve the number sentences and decide which answer is reasonable.

Cross multiply and solve for x:

$11x = (2)(165)$, $11x = 330$, $x = 30$. If there are 165 total cards sold, then 30 of them were birthday cards.

Check your work.

The ratio of birthday cards to total cards is 2:11, so the ratio of actual birthday cards sold to total cards sold should reduce to 2:11. The greatest common factor of 30 and 165 is 15, and 30:165 reduces to 2:11.

11. *Read the entire word problem.*

We are given the ratio of orangutans to gorillas and the total number of orangutans and gorillas.

Identify the question being asked.

We are looking for the number of gorillas.

Underline the keywords and words that indicate formulas.

The word *ratio* means that we will likely be using a ratio to set up a proportion. The word *total* is a signal that this is a part-to-whole ratio problem.

Cross out extra information and translate words into numbers.

There is no extra information in this problem.

List the possible operations.

This problem contains two quantities, orangutans and gorillas. We are given the total of the two quantities, and we are looking for the value of one quantity. This is a part-to-whole problem. Since we are looking for the number of gorillas, we must set up a proportion that compares the ratio of gorillas to the total number of orangutans and gorillas to the ratio of actual orangutans to the total number of actual orangutans and gorillas.

Write number sentences for each operation.

The ratio of orangutans to gorillas is 5:3, which means that the ratio of gorillas to the total number of orangutans and gorillas is 3:(3 + 5), which is 3:8, or $\frac{3}{8}$. Use x to represent the number of gorillas, since that number is unknown. There are 312 total orangutans and gorillas, so the ratio of gorillas to total orangutans and gorillas is x:312, or $\frac{x}{312}$. Set these fractions equal to each other:

$\frac{3}{8} = \frac{x}{312}$

Solve the number sentences and decide which answer is reasonable.

Cross multiply and solve for x:

$8x = (3)(312)$, $8x = 936$, $x = 117$. If there are 312 total cards orangutans and gorillas, then 117 of them are gorillas.

Check your work.

The ratio of gorillas to total orangutans and gorillas is 3:8, so the ratio of actual gorillas to actual total orangutans and gorillas should reduce to 3:8. The greatest common factor of 117 and 312 is 39, and 117:312 reduces to 3:8.

12. *Read the entire word problem.*

We are given the ratio of yearbook writers to newspaper writers and the total number of writers.

Identify the question being asked.

We are looking for the number of yearbook writers.

Underline the keywords and words that indicate formulas.

The word *ratio* means that we will likely be using a ratio to set up a proportion. The word *total* is a signal that this is a part-to-whole ratio problem.

Cross out extra information and translate words into numbers.

There is no extra information in this problem.

List the possible operations.

This problem contains two quantities, yearbook writers and newspaper writers. We are given the total of the two quantities, and we are looking for the value of one quantity. This is a part-to-whole problem. Since we are looking for the number of yearbook writers, we must set up a proportion that compares the ratio of yearbook writers to the total number of writers to the ratio of actual yearbook writers to the total number of actual writers.

Write number sentences for each operation.

The ratio of yearbook writers to newspaper writers is 13:9, which means that the ratio of yearbook writers to the total number of writers is 13:(9 + 13), which is 13:22, or $\frac{13}{22}$. Use x to represent the number of yearbook writers, since that number is unknown. There are 88 total writers, so the ratio of yearbook writers to total writers is x:88, or $\frac{x}{88}$. Set these fractions equal to each other:

$\frac{13}{22} = \frac{x}{88}$

Solve the number sentences and decide which answer is reasonable.

Cross multiply and solve for x:

$22x = (13)(88)$, $22x = 1{,}144$, $x = 52$. If there are 88 total writers, then 52 of them are yearbook writers.

Check your work.

The ratio of yearbook writers to total writers is 13:22, so the ratio of actual yearbook writers to actual total writers should reduce to 13:22. The greatest common factor of 52 and 88 is 4, and 52:88 reduces to 13:22.

13. *Read the entire word problem.*

We are given the ratio of one-way tickets to round-trip tickets and the number of round-trip tickets.

Identify the question being asked.

We are looking for the total number of tickets.

Underline the keywords and words that indicate formulas.

The word *ratio* means that we will likely be using a ratio to set up a proportion. The word *total* is a signal that this is a part-to-whole ratio problem.

Cross out extra information and translate words into numbers.

There is no extra information in this problem.

List the possible operations.

This problem contains two quantities, one-way tickets and round-trip tickets. We are given the value of one quantity, and we are looking for the total of the two quantities. This is a part-to-whole problem. Since we are looking for the total number of tickets, we must set up a proportion that compares the ratio of round-trip tickets to the total number of tickets to the ratio of actual round-trip tickets to the total number of actual tickets.

Write number sentences for each operation.

The ratio of one-way tickets to round-trip tickets is 8:15, which means that the ratio of round-trip tickets to the total number of tickets is

15:(8 + 15), which is 15:23, or $\frac{15}{23}$. Use x to represent the number of total tickets, since that number is unknown. There are 75 round-trip tickets, so the ratio of round-trip tickets to total tickets is 75:x, or $\frac{75}{x}$. Set these fractions equal to each other:

$$\frac{15}{23} = \frac{75}{x}$$

Solve the number sentences and decide which answer is reasonable.

Cross multiply and solve for x:

$15x = (75)(23)$, $15x = 1,725$, $x = 115$. If there were 75 round-trip tickets sold, then there were 115 total tickets sold.

Check your work.

The ratio of round-trip tickets to the total number of tickets is 15:23, so the ratio of actual round-trip tickets to the total number of actual tickets should reduce to 15:23. The greatest common factor of 75 and 115 is 5, and 75:115 reduces to 15:23.

14. *Read the entire word problem.*

We are given the ratio of part-time employees to full-time employees and the number of full-time employees.

Identify the question being asked.

We are looking for the total number of employees.

Underline the keywords and words that indicate formulas.

The word *ratio* means that we will likely be using a ratio to set up a proportion. The word *total* is a signal that this is a part-to-whole ratio problem.

Cross out extra information and translate words into numbers.

There is no extra information in this problem.

List the possible operations.

This problem contains two quantities, part-time employees and full-time employees. We are given the value of one quantity, and we are looking for the total of the two quantities. This is a part-to-whole problem. Since we are looking for the total number of employees, we must set up a proportion that compares the ratio of full-time employees to the total number of employees to the ratio of actual full-time employees to the total number of actual employees.

Write number sentences for each operation.

The ratio of part-time employees to full-time employees is 5:14, which means that the ratio of full-time employees to the total number of employees is 14:(5 + 14), which is 14:19, or $\frac{14}{19}$. Use x to represent the number of total tickets, since that number is unknown. There are 168 full-time employees, so the ratio of full-time employees to total employees is 168:x, or $\frac{168}{x}$. Set these fractions equal to each other:

$$\frac{14}{19} = \frac{168}{x}$$

Solve the number sentences and decide which answer is reasonable.

Cross multiply and solve for x:

$14x = (168)(19)$, $14x = 3,192$, $x = 228$. If there are 168 full-time employees, then there are 228 total employees.

Check your work.

The ratio of full-time employees to the total number of employees is 14:19, so the ratio of actual full-time employees to the total number of actual employees should reduce to 14:19. The greatest common factor of 168 and 228 is 12, and 168:228 reduces to 14:19.

15. *Read the entire word problem.*

We are given the ratio of Scottsdale residents to Wharton residents and the number of Scottsdale residents.

Identify the question being asked.

We are looking for the total number of residents in the two towns.

Underline the keywords and words that indicate formulas.

The word *ratio* means that we will likely be using a ratio to set up a proportion. The word *combined* is a signal that this is a part-to-whole ratio problem.

Cross out extra information and translate words into numbers.

There is no extra information in this problem.

List the possible operations.

This problem contains two quantities, of Scottsdale residents and of Wharton residents. We are given the value of one quantity, and we are looking for the total of the two quantities. This is a part-to-whole problem. Since we are looking for the total number of residents, we must set up a proportion that compares the ratio of Scottsdale residents to the total number of residents to the ratio of actual Scottsdale residents to the total number of actual residents.

Write number sentences for each operation.

The ratio of Scottsdale residents to Wharton residents is 7:12, which means that the ratio of Scottsdale residents to the total number of residents is 7:(12 + 7), which is 7:19, or $\frac{7}{19}$. Use x to represent the number of total residents, since that number is unknown. There are 8,435 Scottsdale residents, so the ratio of Scottsdale residents to total residents is 8,435:x, or $\frac{8,435}{x}$. Set these fractions equal to each other:

$$\frac{7}{19} = \frac{8,435}{x}$$

Solve the number sentences and decide which answer is reasonable.

Cross multiply and solve for x:

$7x = (8,435)(19)$, $7x = 160,265$, $x = 22,895$. If 8,435 people live in Scottsdale, then 160,265 people live in Scottsdale and Wharton combined.

Check your work.

The ratio of Scottsdale residents to the total number of residents is 7:19, so the ratio of actual Scottsdale residents to the total number of actual residents should reduce to 7:19. The greatest common factor of 8,435 and 22,895 is 1,205, and 8,435:22,895 reduces to 7:19.

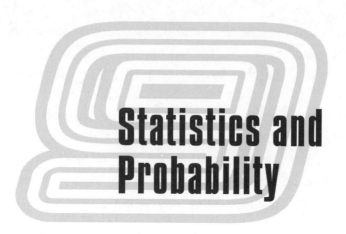

Statistics and Probability

W_hen working with_ a data set, such as a group of test scores or a series of measurements, we can calculate various statistics that help us describe that data set. This data is often presented in a graph, but you may have to calculate statistics based on information in a word problem. In this chapter, we'll look at statistics word problems and probability word problems. As always, we'll use the eight-step process to help us identify the type of problem and the operation or operations needed to solve it.

STATISTICS: SUBJECT REVIEW

A **mean** is an average of a set of values. The mean is found by adding all of the values together and then dividing by the number of values.

A **median** is the middle value of a set after the values in the set are put in order from least to greatest. If there is an even number of values in a set, the median is the average of the two middle values.

A **mode** is the value in a set that occurs the most often. If there are two or more unique values that occur most often, then the set will have more than one mode.

A **range** is the difference between the smallest value and the greatest value in a set.

Given a set of numbers, such as 10, 6, 4, 10, and 30, we can find the mean, median, mode, and range of the set.

The mean of this set is 12, because $10 + 6 + 4 + 10 + 30 = 60$, and $\frac{60}{5}$ $= 12$.

The median of this set is 10, because after the set is ordered, it becomes 4, 6, 10, 10, 30, and the middle value is 10. $10 + 10 = \frac{20}{2} = 10$.

The mode of this set is 10, because it is the value that occurs the most often. There are two 10s in the set, and no other value occurs more than once.

The range of this set is 26, because the lowest value in the set is 4 and the highest value in the set is 30: $30 - 26 = 4$.

Now that we know how to find mean, median, mode, and range, let's look at how to recognize and solve statistics word problems.

MEAN

Some word problems may ask you for the mean of a set, but more often, you'll be asked to find the average. The word *average* is just another word for *mean*, so if a question asks you for an average, it's asking you for a mean.

Example

Nancy plays nine holes of golf. Her scores are 7, 4, 6, 5, 3, 6, 5, 4, and 5. What was her average score for a hole?

Read the entire word problem.
We are given the number of holes of golf Nancy plays and her score on each hole.
Identify the question being asked.
We are looking for her average score for a hole.
Underline the keywords and words that indicate formulas.
The word *average* means that we are looking for the mean of the set.
Cross out extra information and translate words into numbers.
There is no extra information in this problem.
List the possible operations.
The mean is found by adding all of the scores and dividing by the number of scores, so we must use addition and division.
Write number sentences for each operation.

First, find the sum of all the scores. We will need the sum before we can divide:

7 + 4 + 6 + 5 + 3 + 6 + 5 + 4 + 5

Solve the number sentences and decide which answer is reasonable.

7 + 4 + 6 + 5 + 3 + 6 + 5 + 4 + 5 = 45

Write number sentences for each operation.

Now that we have the total of all the scores, divide by the number of scores, 9:

$\frac{45}{9}$

Solve the number sentences and decide which answer is reasonable.

$\frac{45}{9} = 5$

Check your work.

The mean multiplied by the number of values should equal the sum of the values: $5 \times 9 = 45$ and 7 + 4 + 6 + 5 + 3 + 6 + 5 + 4 + 5 = 45, so our answer is correct.

MEDIAN

Again, some word problems may come right out and ask you for the median of a set, but many questions will simply ask you for the middle value. The keyword *middle* should immediately make you think *median*. Less often, you may be asked to find the value at the center of a data set. That's just another way of asking you for the median.

Example

Five students are practicing for the long jump. Karen jumps 75 inches, John jumps 82 inches, Jack jumps 68 inches, Maeve jumps 71 inches, and Clare jumps 72 inches. What is the median length of their jumps?

Read the entire word problem.

We are given the distances that five students jump.

Identify the question being asked.

We are looking for the median length.

Underline the keywords and words that indicate formulas.

The word *median* means that we are looking for the middle value of the set.

Cross out extra information and translate words into numbers.

There is no extra information in this problem.

List the possible operations.

To find the median value, put all of the lengths in order from smallest to greatest and select the middle value.

Write number sentences for each operation.

We don't need a number sentence to solve this kind of problem. We must place the scores in order from smallest to greatest:

68, 71, 72, 75, 82

The middle value is 72, so 72 inches is the median length of their jumps.

Check your work.

We can check that our median is correct by comparing the number of values that are less than or equal to the median to the number of values that are greater than or equal to the median. These numbers should be equal. There are two values that are less than 72 (68 and 71) and two values that are greater than 72 (75 and 82), so 72 is the median.

CAUTION

CHECKING THAT THE number of values that are less than or equal to the median is the same as the number of values that are greater than or equal to the median is not a surefire check. Sometimes, the median is an average of two numbers, and although you may find an answer that is between the two numbers, that does not guarantee that you have divided correctly. Also, if the median value occurs more than once in a set, the number of values that are less than or equal to that median may not be the same as the number of values that are greater than or equal to it. For instance, the set of 1, 2, 3, 3, 5 has a median of 3, and there are two values that are less than or equal to 3 (1 and 2) and three values that are greater than or equal to 3 (3, 3, and 5), but 3 is still the median.

MODE

A word problem may simply ask you for the mode of a set, or more likely, the problem will ask you to find the value that occurs the most. Or it may ask you to find the most common element in a set. The words *most often*, *most frequent*, or *most common* can indicate that you must find the mode.

Example

Bob delivers pizzas after school. His first delivery is 1.2 miles from the pizzeria, his second delivery is 1.8 miles away, his third delivery is 2.3 miles away, his fourth delivery is 1.4 miles away, his fifth delivery is 1.8 miles away, his sixth delivery is 1.4 miles away, his seventh delivery is 2.1 miles away, his eighth delivery is 1.8 miles away, and his last delivery is 1.2 miles away. What is the most common distance Bob has to travel?

Read the entire word problem.
We are given the distances Bob must travel to make each of nine deliveries.
Identify the question being asked.
We are looking for the distance he travels most often.
Underline the keywords and words that indicate formulas.
The words *most common* mean that we are looking for the mode of the set.
Cross out extra information and translate words into numbers.
There is no extra information in this problem.
List the possible operations.

The mode is found by counting the number of times each value occurs in a set.

Write number sentences for each operation.

We don't need a number sentence to solve this kind of problem. The distances 2.1 miles and 2.3 miles each occur once in the set, the distances 1.2 miles and 1.4 miles occur twice each in the set, and the distance 1.8 miles occurs three times in the set. Since the distance 1.8 miles occurs more often than any other distance, 1.8 miles is the mode of the set and the most common distance Bob has to travel.

Check your work.

The value 1.8 miles occurs three times in the set. Check that no other value occurs more than twice in the set. The values 1.4 and 1.2 occur twice each, and 2.1 and 2.3 occur once each, so 1.8 is indeed the mode of the set because it occurs the most often.

Finding the mode of a set given a bar graph, chart, table, or plot can be easy, but counting unique values given in a word problem can be difficult. When trying to solve a mode word problem, construct a frequency table. A frequency table has three columns: one column to note each unique value, a column to make tally marks every time those values appear in the set, and a column to total your tallies when you're done. The value with the highest total must be the mode of the set.

Let's try that last problem again with a frequency table. As you encounter a new value in the set, add a row to the table, and after every value you encounter, place a tally mark in the table. The first value given is 1.2 miles. Add it to the table and place a tally next to it:

Value	Tally	Total
1.2	I	

The next value is 1.8. Add it to the table, and place a tally next to it:

Value	Tally	Total
1.2	I	
1.8	I	

The next value is 2.3, followed by 1.4. Add those values and tallies to the table:

Value	Tally	Total
1.2	|	
1.8	|	
2.3	|	
1.4	|	

The next value, 1.8, already exists in the table, so add another tally in that row:

Value	Tally	Total
1.2	|	
1.8	||	
2.3	|	
1.4	|	

Continue until all of the values have been added and tallied:

Value	Tally	Total
1.2	||	
1.8	|||	
2.3	|	
1.4	||	
2.1		

Finally, add the tallies to find the total for each value:

Value	Tally	Total
1.2	||	1
1.8	|||	3
2.3	|	1
1.4	||	2
2.1	|	1

It's easy to see now that the value 1.8 occurs more often than the other values, so 1.8 miles is the mode of the set.

RANGE

The range of a data set is the difference between the smallest and largest values of the set, so a word problem may ask you for exactly that: the difference between the smallest and largest values of the set. Or it may ask you for the *span* or *reach* of the data set. These keywords tell you to find the range of the data set.

Example

Meghan has eight sunflowers. She uses a meter stick to find the heights of the flowers: 1.3 meters, 0.8 meters, 1.2 meters, 1.7 meters, 0.8 meters, 0.9 meters, 1.1 meters, and 1.5 meters. The heights of her sunflowers span how many meters?

Read the entire word problem.
We are given the heights of eight sunflowers.
Identify the question being asked.
We are looking for the number of meters that these heights span.
Underline the keywords and words that indicate formulas.
The word *span* means that we are looking for the range of the set.
Cross out extra information and translate words into numbers.
The number of sunflowers, 8, is not needed to solve this problem, so that number can be crossed out.
List the possible operations.
The range is found by subtracting the smallest value from the largest value.
Write number sentences for each operation.
The smallest number in the set is 0.8 meters, and the largest number in the set is 1.7 meters:
1.7 − 0.8
Solve the number sentences and decide which answer is reasonable.
1.7 − 0.8 = 0.9 meters
Check your work.

Since we used subtraction to find our answer, we can use addition to check our answer. The smallest value plus the range should equal the largest value: $0.8 + 0.9 = 1.7$, which is the largest value in the set.

DIRECTIONS: Use the following information to answer questions 1–4.

Joseph plays a video game ten times and scores the following points each time: 320, 285, 300, 290, 320, 310, 300, 305, 330, and 320.

1. How many points did Joseph score most often?

2. How many points did Joseph score on average?

3. What is the range of Joseph's scores?

4. What was the middle point total of Joseph's scores?

5. Justin visits five parks in his town. Shady Park has eight benches, Hillside Park has 13 benches, Lakeville Park has ten benches, and Brighton Park has eight benches. If the mean number of benches is ten, how many benches are in the last park, Long Park?

6. Eric sells pretzels at the beach from 9 A.M. to 4 P.M. He sells 16 pretzels, 19 pretzels, 28 pretzels, 34 pretzels, 39 pretzels, 28 pretzels, and 20 pretzels each hour, respectively. How many pretzels span the difference from Eric's most successful hour to his least successful hour?

7. Katie's test scores are 85, 95, 100, 85, 90, 95, 85, 95, and 85. If Katie can drop her two lowest test scores, what will be her most common test score?

8. Annie buys four tomatoes, three heads of lettuce, eight peppers, 16 carrots, three onions, and eight potatoes. If the median number of vegetables she buys is 4, what are all the possible numbers of eggplant Annie may have bought?

MEASURE THE HEIGHT of everyone in your family. What is the range of heights in your family? What is the average height and median height? Is there a height that occurs more than once?

PROBABILITY: SUBJECT REVIEW

If you've ever wondered what the chances are that a certain something may happen, then you've thought about probability. Unlike many fraction and decimal questions, which are typically straight adding, subtracting, multiplying, or dividing, or statistic questions, which typically come from reading a table or graph, probability questions typically *are* word problems.

Probability is the likelihood that an event or events will occur, usually given as a fraction in which the numerator is the number of possibilities that allow for the event to occur and the denominator is the total number of possibilities.

To find the probability that a single event will occur, place the number of possibilities that allow the event to occur over the number of total possibilities. To find the probability that one of multiple events will occur, add the probabilities of those events. To find the probability that two or more events will occur, multiply the probabilities of those events.

Probability word problems almost always directly ask "what is the probability" of an event occurring, but the phrases *what are the chances*, *what are the odds*, and *what is the likelihood* can also signal a probability question. The words *at random* often indicate a probability question, too.

Simple probability questions do not involve any operations or number sentence, but more complex questions may require you to add or multiply. If a probability question contains the word *or*, you will likely have to add more than one fraction to find the probability of the event(s) occurring. If a probability question contains the word *and*, you will likely have to multiply two fractions to find the probability that both events will occur. If a probability question contains the word *remove*, you will likely have not only to multiply, but to change at least the denominator of one probability, because an element has been removed from a data set.

Example

> A spinner is divided into five equal slices, numbered one through
> five. If Noreen spins the spinner, what is the likelihood that it will
> land on the number two?

As always, we will use the eight-step process to solve this word problem.

Read the entire word problem.

We are told that a spinner is divided into five equal slices, numbered
one through five, and that the spinner is spun once.

Identify the question being asked.

We are looking for the likelihood that it lands on the number two.

Underline the keywords and words that indicate formulas.

The word *likelihood* means that we are looking for a probability.

Cross out extra information and translate words into numbers.

There is no extra information in this problem.

List the possible operations.

To find the probability, we must find the total number of possibilities
(the total number of results that can come from spinning the spinner) and the number of those possibilities that make the event (the
spinner landing on the number two) true.

Write number sentences for each operation.

We do not need to add or multiply any probabilities, but we do need
to write a fraction. The spinner has five slices, so the total number of
possibilities that can result from one spin is five. Only one slice is
numbered two, so there is only one slice that can make the event
true. The probability of Noreen spinning a two is $\frac{1}{5}$.

Check your work.

There is no formal way to check a probability answer. Read the question again and check that you correctly found the total number of
possibilities and the number of outcomes that make the event true.

Example

> Noreen spins the same spinner again. What is the probability that
> it lands on a number that is greater than three?

Read the entire word problem.

We are told that a spinner is divided into five equal slices, numbered one through five, and that the spinner is spun once.

Identify the question being asked.

We are looking for the likelihood that it lands on a number that is greater than three.

Underline the keywords and words that indicate formulas.

The word *probability* tells us that we are looking for a probability.

Cross out extra information and translate words into numbers.

There is no extra information in this problem.

List the possible operations.

The total number of possibilities is five again, but the number of outcomes that makes the event true has changed. We must find how many numbers there are on the spinner that are greater than three.

Write number sentences for each operation.

The numbers four and five are greater than three, so there are two slices that the spinner could land on that would make the event true. The probability of Noreen spinning a number greater than three is $\frac{2}{5}$.

- PRACTICE LAP -

DIRECTIONS: Use the following information to answer the questions that follow.

Three sides of a cube are painted red, one side of the cube is painted yellow, and two sides of the cube are painted blue. The cube is rolled once.

9. What are the chances the cube will land on a red side?

10. What is the probability that the cube will land on the yellow side?

11. What are the odds that the cube will land on a blue side?

12. What is the likelihood that the cube will NOT land on yellow?

Example

A jar contains five pennies, nine nickels, six dimes, and four quarters. If Marty selects one coin at random, what is the probability that the coin is either a penny or a nickel?

Read the entire word problem.

We are told that a jar contains five pennies, nine nickels, six dimes, and four quarters and that one coin is selected.

Identify the question being asked.

We are looking for the probability that it is a penny or a nickel.

Underline the keywords and words that indicate formulas.

The word *probability* tells us that we are looking for a probability. The word *or* tells us that we are looking for two probabilities, which we will likely have to add.

Cross out extra information and translate words into numbers.

There is no extra information in this problem.

List the possible operations.

The total number of possibilities is the total number of coins in the jar, so we must first find that total.

Write number sentences for each operation.

$5 + 9 + 6 + 4$

Solve the number sentences and decide which answer is reasonable.

$5 + 9 + 6 + 4 = 24$

Write number sentences for each operation.

Since there are five pennies in the jar, the probability of selecting a penny is $\frac{5}{24}$. There are nine nickels in the jar, so the probability of selecting a nickel is $\frac{9}{24}$. To find the probability that a penny or nickel is selected, add the two probabilities:

$\frac{5}{24} + \frac{9}{24}$

Solve the number sentences and decide which answer is reasonable.

$\frac{5}{24} + \frac{9}{24} = \frac{14}{24}$

Always reduce fractions to simplest form. The probability of Marty selecting either a penny or a nickel is $\frac{14}{24}$, or $\frac{7}{12}$.

CAUTION

WHEN WORKING WITH probabilities, never reduce fractions until you have added. We could have reduced the probability of selecting a nickel, $\frac{9}{24}$, but then we would not have been able to add it to the probability to selecting a penny. Always complete any addition of probabilities before reducing.

PRACTICE LAP

13. Al has eight white shirts, five blue shirts, two yellow shirts, and one red shirt in his closet. If he chooses one shirt at random, what are the chances that the shirt will be blue or red?

14. Ingrid has 24 rock songs, 40 pop songs, ten classical songs, and six jazz songs on her portable music player. If the player selects a song at random, what is the probability it will be a rock song or a jazz song?

15. A gumball machine contains 38 lemon gumballs, 12 grape gumballs, 19 strawberry gumballs, 16 orange gumballs, and 15 blueberry gumballs. If Mohammed pulls a gumball from the machine at random, what are the chances it will be lemon, strawberry, or orange?

16. There are 400 orchestra seats at Hampton Hall, 300 seats in the first mezzanine, 200 seats in the second mezzanine, 100 seats in the third mezzanine, and 100 seats in the fourth mezzanine. If Kayla wins a ticket to a concert at the hall, what are the odds the ticket is NOT in the orchestra?

Sometimes, a problem asks us to find the probability of two or more events occurring. That's when we need to use multiplication.

Example

A deck contains 15 cards, three of which are blue, five of which are red, two of which are black, and five of which are green. Phil selects a card from the deck, writes down the color, and then inserts it back into the deck. He selects a second card and writes down the color of the second card. What is the probability that Phil selected a red card and then a black card?

Read the entire word problem.
We are told that a deck contains 15 cards, three of which are blue, five of which are red, two of which are black, and five of which are green.
Identify the question being asked.
We are looking for the probability that the two cards selected were red and black, respectively.
Underline the keywords and words that indicate formulas.
The word *probability* tells us that we are looking for a probability, and the keyword *and* tells us that we are looking for two probabilities, which we will likely have to multiply.
Cross out extra information and translate words into numbers.
There is no extra information in this problem.
List the possible operations.
The total number of possibilities is the total number of cards in the deck, which we are told is 15. Phil selects a red card first. Since there are 15 cards in the deck and five of them are red, the probability of him selecting a red card is $\frac{5}{15}$. Phil selects a black card next. Since there are 15 cards in the deck and two of them are red, the probability of him selecting a black card is $\frac{2}{15}$.
Write number sentences for each operation.
The probability that Phil selects a red card and then a black card is the product of each of those probabilities:
$\frac{5}{15} \times \frac{2}{15}$
Solve the number sentences and decide which answer is reasonable.
$\frac{5}{15} \times \frac{2}{15} = \frac{10}{225} = \frac{2}{45}$
The probability that Phil selects a red card and then a black card is $\frac{2}{45}$.

INSIDE TRACK

IT'S BEST NOT to reduce fractions before adding them, because you may lose your common denominators, but it is always best to reduce fractions before multiplying. In the last example, we could have reduced $\frac{5}{15}$ to $\frac{1}{3}$, and that would have made multiplying by $\frac{2}{15}$ easier. We also would not have needed to reduce our answer—and it's a lot easier to reduce $\frac{5}{15}$ than $\frac{10}{225}$!

When working with multiple probabilities, we must always be careful of removals. In the last example, Phil selected a card, wrote it down, and then returned the card to the deck. So, when Phil selected his second card, the probability was still a number over 15, since there were still 15 cards in the deck. However, sometimes a word problem will describe a similar situation, but state that the card is *removed* from the deck after it is selected. If Phil had removed the first card from the deck and not returned it, the probability of selecting a black card would have been $\frac{2}{14}$ since there would have been only 14 cards in the deck. Read carefully to see if any objects are removed from multiple probability problems—the numerators and denominators of your probabilities may change. Let's look at an example similar to the last, but with a removal.

Example

A deck contains 15 cards, three of which are blue, five of which are red, two of which are black, and five of which are green. Phil selects a card from the deck, removes it, and then selects another card. What is the likelihood that Phil selected two green cards?

Read the entire word problem.

Again, we are told that a deck contains 15 cards, three of which are blue, five of which are red, two of which are black, and five of which are green.

Identify the question being asked.

We are looking for the probability that two green cards are selected, but we are also told that the first card was not returned to the deck after it was selected.

Underline the keywords and words that indicate formulas.

The word *probability* tells us that we are looking for a probability, and the keyword *and* tells us that we are looking for two probabilities, which we will likely have to multiply. The keyword *remove* tells us that the denominator of our second probability will not be the same as the denominator of our first probability and that the numerator may change as well.

Cross out extra information and translate words into numbers.

There is no extra information in this problem.

List the possible operations.

The total number of possibilities is the total number of cards in the deck at any time. To start, there are 15 cards in the deck. Since there are five green cards in the deck, the probability of him selecting a green card is $\frac{5}{15}$. If Phil selects and removes a green card, there will be only 14 cards left in the deck, four of which are green. The probability that he will select a green card the second time is $\frac{4}{14}$.

Write number sentences for each operation.

The probability that Phil selects two green cards is the product of each of those probabilities:

$\frac{5}{15} \times \frac{4}{14}$

Reduce the fractions before multiplying: $\frac{5}{15} = \frac{1}{3}$ and $\frac{4}{14} = \frac{2}{7}$.

Solve the number sentences and decide which answer is reasonable.

$\frac{1}{3} \times \frac{2}{7} = \frac{2}{21}$

The probability that Phil selects a green card, removes it from the deck, and then selects another green card is $\frac{2}{21}$.

17. There are 25 students in Mr. Scott's class. Only one perfect score was recorded for the math exam, and only one perfect score was recorded for the science exam. What are the odds that Stephanie, a student in the class, received both perfect scores?

18. A spinner is divided into ten slices, numbered one through ten, respectively. If Orla spins the spinner twice, what is the probability that it will land on five the first time and a number greater than five the second time?

19. Hayley rolls a number cube three times. What are the chances that it lands on an even number the first time, an odd number the second time, and the number six the third time?

20. A package contains 12 chocolate cookies and ten vanilla cookies. If Rita takes a cookie at random and eats it, and then takes another cookie from the package, what is the probability that both cookies were vanilla cookies?

21. An after-school program offers five science courses, eight math courses, four art courses, and three music courses. If Mairead is placed in two courses at random, what are the odds that she is placed in an art course and a music course? (Note: She cannot be placed in the same course twice.)

22. Molly and Anna share a sock drawer which contains eight pairs of brown socks, 18 pairs of white socks, and 20 pairs of black socks. If each girl selects a pair of socks from the drawer to wear, what is the likelihood that Molly selected a white pair and Anna selected a black pair?

PACE YOURSELF

TAKE A NUMBER cube and cover it with paper. Write the letters *A*, *B*, *C*, *D*, *E*, and *F* on the cube, with one letter on each side. If you were to roll the cube five times, what is the probability that it would land on *A* every time? Roll the cube and record your results. What is the probability if you rolled the cube twice that it would land on a vowel both times? What if you replaced the vowel with an *X* after the cube landed on a vowel—what would be the probability of rolling a vowel twice in a row?

SUMMARY

IN THIS CHAPTER, we learned what keywords indicate that a word problem is asking for mean, median, mode, or range. We also learned how to spot not just probability word problems, but probability word problems that require addition and multiplication. Up next: geometry word problems.

ANSWERS

1. *Read the entire word problem.*
   We are given Joseph's ten scores.
   Identify the question being asked.
   We are looking for the number of points he scored most often.
   Underline the keywords and words that indicate formulas.
   The words *most often* mean that we are looking for the mode of the set.
   Cross out extra information and translate words into numbers.
   The number of times Joseph played the game is not needed to solve this problem, so the number ten can be crossed out.
   List the possible operations.

The mode is found by counting the number of times each value occurs in a set.

Write number sentences for each operation.

We don't need a number sentence to solve this kind of problem. The scores 285, 290, 305, 310, and 330 occur once each. The score 300 occurs twice, and the score 320 occurs three times, which means that 320 is the mode.

Check your work.

The value 320 occurs three times in the set. No other value occurs more than twice, so 320 must be the mode.

2. *Read the entire word problem.*

We are given Joseph's ten scores.

Identify the question being asked.

We are looking for the number of points he scored on average.

Underline the keywords and words that indicate formulas.

The word *average* means that we are looking for the mean of the set.

Cross out extra information and translate words into numbers.

There is no extra information in this problem.

List the possible operations.

The mean is found by adding all of the scores and dividing by the number of scores, so we must use addition and division.

Write number sentences for each operation.

First, find the sum of all the scores. We will need the sum before we can divide:

$320 + 285 + 300 + 290 + 320 + 310 + 300 + 305 + 330 + 320$

Solve the number sentences and decide which answer is reasonable.

$320 + 285 + 300 + 290 + 320 + 310 + 300 + 305 + 330 + 320 = 3,080$

Write number sentences for each operation.

Now that we have the total of all the scores, divide by the number of scores, 10:

$$\frac{3,080}{10}$$

Solve the number sentences and decide which answer is reasonable.

$$\frac{3,080}{10} = 308$$

Check your work.

The mean multiplied by the number of values should equal the sum of the values: $308 \times 10 = 3{,}080$ and $320 + 285 + 300 + 290 + 320 + 310 + 300 + 305 + 330 + 320 = 3{,}080$, so 308 must be the mean.

3. *Read the entire word problem.*

   We are given Joseph's ten scores.

   Identify the question being asked.

   We are looking for the range of his scores.

   Underline the keywords and words that indicate formulas.

   The word *range* actually appears in the problem, so we are told that we are looking for a range.

   Cross out extra information and translate words into numbers.

   The number of times Joseph played the game is not needed to solve this problem, so that number can be crossed out.

   List the possible operations.

   The range is found by subtracting the smallest value from the largest value.

   Write number sentences for each operation.

   The smallest number in the set is 285 and the largest number in the set is 330:

   $330 - 285$

   Solve the number sentences and decide which answer is reasonable.

   $330 - 285 = 45$

   Check your work.

   Since we used subtraction to find our answer, we can use addition to check our work. The smallest value plus the range should equal the largest value: $285 + 45 = 330$, which is the largest value in the set.

4. *Read the entire word problem.*

   We are given Joseph's ten scores.

   Identify the question being asked.

   We are looking for the middle value of his scores.

   Underline the keywords and words that indicate formulas.

   The word *middle* means that we are looking for the median value of the set.

   Cross out extra information and translate words into numbers.

   There is no extra information in this problem.

   List the possible operations.

To find the median value, put all of the lengths in order from smallest to greatest and select the middle value. Since there is an even number of scores, the median value will be the average of the fifth and sixth scores.

Write number sentences for each operation.

First, we must place the scores in order from smallest to greatest:

285, 290, 300, 300, 305, 310, 320, 320, 320, 330

The middle values are 305 and 310. The average of these numbers is equal to their sum divided by 2, so first we must find the sum:

305 + 310

Solve the number sentences and decide which answer is reasonable.

305 + 310 = 615

Write number sentences for each operation.

Now, divide that sum by 2:

$\frac{615}{2}$

Solve the number sentences and decide which answer is reasonable.

$\frac{615}{2} = 307.5$

Check your work.

We can check that our median is correct by comparing the number of values that are less than or equal to the median to the number of values that are greater than or equal to the median. These numbers should be equal. There are five values that are less than 307.5 and five values that are greater than 307.5, so 307.5 is the median.

5. *Read the entire word problem.*

We are told that Justin visits five parks, we are given the number of benches in four of those parks, and we are given the mean number of benches in the parks.

Identify the question being asked.

We are looking for the number of benches in Long Park.

Underline the keywords and words that indicate formulas.

The word *mean* usually indicates that we are looking for the mean value of the set, but we are given the mean, and we must use it to find a value in the set.

Cross out extra information and translate words into numbers.

There is no extra information in this problem.

List the possible operations.

Normally, we find the mean by adding all of the values in a set and dividing by the number of values. Since we are given all but one of the values and the mean, we must work backward. First, multiply the mean by the number of values. Then, find the sum of the values and subtract it from that product to find the missing value.

Write number sentences for each operation.

Multiply the mean by the number of values:

10×5

Solve the number sentences and decide which answer is reasonable.

$10 \times 5 = 50$

Write number sentences for each operation.

Next, find the total of the four given values:

$8 + 13 + 10 + 8$

Solve the number sentences and decide which answer is reasonable.

$8 + 13 + 10 + 8 = 39$

Write number sentences for each operation.

Finally, subtract the number of benches in four parks from the product of the mean and the number of parks:

$50 - 39$

Solve the number sentences and decide which answer is reasonable.

$50 - 39 = 11$

There are 11 benches in Long Park.

Check your work.

If there are 11 benches in Long Park, then the sum of the benches in the five parks divided by five should equal ten, since that is the mean number of benches: $8 + 13 + 10 + 8 + 11 = 50$, $\frac{50}{5} = 10$ benches.

6. *Read the entire word problem.*

We are given the number of pretzels Eric sells in each of seven hours.

Identify the question being asked.

We are looking for the difference between the greatest number of pretzels sold in an hour and the least number of pretzels sold in an hour.

Underline the keywords and words that indicate formulas.

The word *span* indicates that we are looking for the range of the set.

Cross out extra information and translate words into numbers.

The hours that Eric worked, 9 A.M. to 4 P.M., are not needed to solve the problem, so that information can be crossed out.

List the possible operations.

The range is found by subtracting the smallest value from the largest value. In this set, the largest value is 39 and the smallest value is 16.

Write number sentences for each operation.

39 – 16

Solve the number sentences and decide which answer is reasonable.

39 – 16 = 25

The difference from Eric's most successful hour to his least successful hour is 25 pretzels.

Check your work.

The smallest value plus the range should equal the largest value: 16 + 25 = 39, which is the largest value in the set.

7. *Read the entire word problem.*

We are given nine test scores for Katie and told that she can drop the two lowest.

Identify the question being asked.

We are looking for her most common test score after dropping those two scores.

Underline the keywords and words that indicate formulas.

The words *most common* indicate that we are looking for the mode of the set.

Cross out extra information and translate words into numbers.

There is no extra information in this problem.

List the possible operations.

We must put the set of scores in order:

85, 85, 85, 85, 90, 95, 95, 95, 100

and remove the two lowest scores:

85, 85, 90, 95, 95, 95, 100

Write number sentences for each operation.

There are no number sentences to write. The score 95 occurs three times, the score 85 occurs twice, and the scores 90 and 100 each occur once. Since 95 occurs the most often, it is the mode of the set and is Katie's most common test score.

Check your work.

After removing the two lowest scores, check that every other score occurs fewer than three times. Since 85 is the only score that occurs more than once, and it occurs only twice, 95 is the most common test score.

8. *Read the entire word problem.*

   We are given the amounts of six different vegetables that Annie buys and the median number of vegetables that she buys.

   Identify the question being asked.

   We are looking for the possible numbers of eggplant that she may have bought.

   Underline the keywords and words that indicate formulas.

   The word *median* usually indicates that we are looking for the middle value of a set, but we are given the median value in this problem. We must use the median to find the missing value in the set.

   Cross out extra information and translate words into numbers.

   There is no extra information in this problem.

   List the possible operations.

   We must put the numbers of vegetables Annie bought in order:

   3, 3, 4, 8, 8, 16

   There are no operations to perform, but we know that the median number of vegetables Annie buys is 4. There are six values in the data set now, and if she buys four eggplants, 4 would be the middle number of the set:

   3, 3, 4, 4, 8, 8, 16

   However, if Annie buys more than four eggplants, then 4 would not be the median of the set, because there would be four values greater than 4 and only two values less than 4. If Annie buys fewer than four eggplants, 4 will also be the median of the set, since there will be three values less than 4 and three values greater than 4. If Annie buys 0, 1, 2, 3, or 4 egg-plants, then the median number of vegetables she buys will be 4.

   Check your work.

   Insert each of those five answers into the set to see if the median value is 4 each time:

   0, 3, 3, **4**, 8, 8, 16

   1, 3, 3, **4**, 8, 8, 16

   2, 3, 3, **4**, 8, 8, 16

   3, 3, 3, **4**, 8, 8, 16

   3, 3, 4, **4**, 8, 8, 16

In each of these sets, 4 is the median value, so all of these answers are correct. If we insert any number other than these five, the median will not be 4.

9. *Read the entire word problem.*

We are told that three sides of cube are red, one side is yellow, and two sides are blue and that the cube is rolled once.

Identify the question being asked.

We are looking for the chances that it will land on a red side.

Underline the keywords and words that indicate formulas.

The word *chances* tells us that we are looking for a probability.

Cross out extra information and translate words into numbers.

There is no extra information in this problem.

List the possible operations.

A cube has six sides, so the total number of possibilities is six. There are three sides that are red.

Write number sentences for each operation.

A number sentence is not needed for this probability. The probability is equal to the number of outcomes that make the event true, 3, over the total number of possibilities, 6: $\frac{3}{6}$, which reduces to $\frac{1}{2}$.

10. *Read the entire word problem.*

We are told that three sides of cube are red, one side is yellow, and two sides are blue and that the cube is rolled once.

Identify the question being asked.

We are looking for the probability that it will land on the yellow side.

Underline the keywords and words that indicate formulas.

The word *probability* tells us that we are looking for a probability.

Cross out extra information and translate words into numbers.

There is no extra information in this problem.

List the possible operations.

A cube has six sides, so the total number of possibilities is six. There is one side that is yellow.

Write number sentences for each operation.

A number sentence is not needed for this probability. The probability is equal to the number of outcomes that make the event true, 1, over the total number of possibilities, 6: $\frac{1}{6}$.

11. *Read the entire word problem.*

    We are told that three sides of the cube are red, one side is yellow, and two sides are blue and that the cube is rolled once.

    Identify the question being asked.

    We are looking for the odds that it will land on a blue side.

    Underline the keywords and words that indicate formulas.

    The word *odds* tells us that we are looking for a probability.

    Cross out extra information and translate words into numbers.

    There is no extra information in this problem.

    List the possible operations.

    A cube has six sides, so the total number of possibilities is six. There are two sides that are blue.

    Write number sentences for each operation.

    A number sentence is not needed for this probability. The probability is equal to the number of outcomes that make the event true, 2, over the total number of possibilities, 6: $\frac{2}{6}$, which reduces to $\frac{1}{3}$.

12. *Read the entire word problem.*

    We are told that three sides of the cube are red, one side is yellow, and two sides are blue and that the cube is rolled once.

    Identify the question being asked.

    We are looking for the likelihood that it will not land on yellow.

    Underline the keywords and words that indicate formulas.

    The word *likelihood* tells us that we are looking for a probability.

    Cross out extra information and translate words into numbers.

    There is no extra information in this problem.

    List the possible operations.

    A cube has six sides, so the total number of possibilities is six. There are three sides that are red and two sides that are blue.

    Write number sentences for each operation.

    Add the number of red sides to the number of blue sides:

    $3 + 2$

    Solve the number sentences and decide which answer is reasonable.

    $3 + 2 = 5$

    The probability is equal to the number of outcomes that make the event true, 5, over the total number of possibilities, 6: $\frac{5}{6}$.

13. *Read the entire word problem.*

We are told that Al has eight white shirts, five blue shirts, two yellow shirts, and one red shirt in his closet.

Identify the question being asked.

We are looking for the chances that he will select a shirt that is blue or red.

Underline the keywords and words that indicate formulas.

The word *chances* tells us that we are looking for a probability. The word *or* tells us that we are looking for two probabilities, which we will likely have to add.

Cross out extra information and translate words into numbers.

There is no extra information in this problem.

List the possible operations.

The total number of possibilities is the total number of shirts in the closet, so we must first find that total.

Write number sentences for each operation.

$8 + 5 + 2 + 1$

Solve the number sentences and decide which answer is reasonable.

$8 + 5 + 2 + 1 = 16$

Write number sentences for each operation.

Since there are five blue shirts in the closet, the probability of selecting a blue shirt is $\frac{5}{16}$. There is one red shirt in the closet, so the probability of selecting a red shirt is $\frac{1}{16}$. To find the probability that a blue shirt or red shirt is selected, add the two probabilities:

$\frac{5}{16} + \frac{1}{16}$

Solve the number sentences and decide which answer is reasonable.

$\frac{5}{16} + \frac{1}{16} = \frac{6}{16}$, or $\frac{3}{8}$

The probability of Al selecting either a blue shirt or a red shirt is $\frac{3}{8}$.

14. *Read the entire word problem.*

We are told that Ingrid has 24 rock songs, 40 pop songs, ten classical songs, and six jazz songs.

Identify the question being asked.

We are looking for the probability that the player will select a rock song or a jazz song.

Underline the keywords and words that indicate formulas.

The word *probability* tells us that we are looking for a probability. The word *or* tells us that we are looking for two probabilities, which we will likely have to add.

Cross out extra information and translate words into numbers.

There is no extra information in this problem.

List the possible operations.

The total number of possibilities is the total number of songs on the player, so we must first find that total.

Write number sentences for each operation.

24 + 40 + 10 + 6

Solve the number sentences and decide which answer is reasonable.

24 + 40 + 10 + 6 = 80

Write number sentences for each operation.

Since there are 24 rock songs on the player, the probability of it selecting a rock song is $\frac{24}{80}$. There are six jazz songs on the player, so the probability of it selecting a jazz song is $\frac{6}{80}$. To find the probability that a rock song or a jazz song is selected, add the two probabilities:

$\frac{24}{80} + \frac{6}{80}$

Solve the number sentences and decide which answer is reasonable.

$\frac{24}{80} + \frac{6}{80} = \frac{30}{80}$, or $\frac{3}{8}$

The probability of the player selecting either a rock song or a jazz song is $\frac{3}{8}$.

15. *Read the entire word problem.*

We are told that a gumball machine contains 38 lemon gumballs, 12 grape gumballs, 19 strawberry gumballs, 16 orange gumballs, and 15 blueberry gumballs.

Identify the question being asked.

We are looking for the chances that Mohammed will select a lemon, strawberry, or orange gumball.

Underline the keywords and words that indicate formulas.

The word *chances* tells us that we are looking for a probability. The word *or* tells us that we are looking for more than one probability, which we will likely have to add.

Cross out extra information and translate words into numbers.

There is no extra information in this problem.

List the possible operations.

The total number of possibilities is the total number of gumballs in the machine, so we must first find that total.

Write number sentences for each operation.

38 + 12 + 19 + 16 + 15

Solve the number sentences and decide which answer is reasonable.

38 + 12 + 19 + 16 + 15 = 100

Write number sentences for each operation.

Since there are 38 lemon gumballs, the probability of Mohammed selecting a lemon gumball is $\frac{38}{100}$. There are 19 strawberry gumballs, so the probability of him selecting a strawberry gumball is $\frac{19}{100}$, and since there are 16 orange gumballs, the probability of him selecting an orange gumball is $\frac{16}{100}$. To find the probability that a lemon, strawberry, or orange gumball is selected, add the three probabilities:

$\frac{38}{100} + \frac{19}{100} + \frac{16}{100}$

Solve the number sentences and decide which answer is reasonable.

$\frac{38}{100} + \frac{19}{100} + \frac{16}{100} = \frac{73}{100}$

The probability of Mohammed selecting a lemon, strawberry, or orange gumball is $\frac{73}{100}$.

16. *Read the entire word problem.*

We are told that there are 400 orchestra seats, 300 first mezzanine seats, 200 second mezzanine seats, 100 third mezzanine seats, and 100 fourth mezzanine seats.

Identify the question being asked.

We are looking for the odds that Kayla will not win an orchestra ticket.

Underline the keywords and words that indicate formulas.

The word *odds* tells us that we are looking for a probability.

Cross out extra information and translate words into numbers.

There is no extra information in this problem.

List the possible operations.

The total number of possibilities is the total number of tickets in the hall, so we must first find that total.

Write number sentences for each operation.

400 + 300 + 200 + 100 + 100

Solve the number sentences and decide which answer is reasonable.

400 + 300 + 200 + 100 + 100 = 1,100

Write number sentences for each operation.

The probability of Kayla winning a ticket in the first mezzanine is $\frac{300}{1,100}$, since there are 300 seats in that section out of a total of 1,100 seats. In the same way, the probability of her winning a ticket in the second mezzanine is $\frac{200}{1,100}$, the probability of her winning a ticket in the third mezzanine is $\frac{100}{1,100}$, and the probability of her winning a ticket in the fourth mezzanine is $\frac{100}{1,100}$. To find the probability that she won a ticket in one of these sections, add the four probabilities:

$$\frac{300}{1,100} + \frac{200}{1,100} + \frac{100}{1,100} + \frac{100}{1,100}$$

Solve the number sentences and decide which answer is reasonable.

$$\frac{300}{1,100} + \frac{200}{1,100} + \frac{100}{1,100} + \frac{100}{1,100} = \frac{700}{1,100} = \frac{7}{11}$$

The odds of Kayla winning a ticket that is not in the orchestra is $\frac{7}{11}$.

17. *Read the entire word problem.*

We are given the number of students in Mr. Scott's class and told that only one perfect score was received on each of two exams.

Identify the question being asked.

We are looking for the odds that Stephanie has received both scores.

Underline the keywords and words that indicate formulas.

The word *odds* tells us that we are looking for a probability, and the keyword *and* tells us that we are looking for two probabilities, which we will likely have to multiply.

Cross out extra information and translate words into numbers.

There is no extra information in this problem.

List the possible operations.

The total number of possibilities is the total number of students in the class, which we are told is 25. Stephanie is only one student, so the probability that she has received a perfect score on the math exam is $\frac{1}{25}$. In the same way, the probability that she received a perfect score on the science exam is also $\frac{1}{25}$.

Write number sentences for each operation.

The probability that Stephanie is the student to receive a perfect score on both exams is the product of each of those probabilities:

$$\frac{1}{25} \times \frac{1}{25}$$

Solve the number sentences and decide which answer is reasonable.

$$\frac{1}{25} \times \frac{1}{25} = \frac{1}{625}$$

The probability that Stephanie was the student to receive a perfect score on both exams is $\frac{1}{625}$.

18. *Read the entire word problem.*

We are told that a spinner is divided into ten slices, numbered one through ten, respectively, and that the spinner is spun twice.

Identify the question being asked.

We are looking for the probability that it will land on five the first time and a number greater than five the second time.

Underline the keywords and words that indicate formulas.

The word *probability* tells us that we are looking for a probability, and the keyword *and* tells us that we are looking for two probabilities, which we will likely have to multiply.

Cross out extra information and translate words into numbers.

There is no extra information in this problem.

List the possible operations.

The total number of possibilities is the total number of slices on the spinner, which we are told is ten. Since only one of the ten slices is numbered five, the probability of the spinner landing on five is $\frac{1}{10}$. There are five slices that contain numbers that are greater than five (six, seven, eight, nine, ten), so the probability of the spinner landing on a number greater than five the second time is $\frac{5}{10}$, or $\frac{1}{2}$.

Write number sentences for each operation.

The probability that the spinner lands on five and then on a number greater than five is the product of each of those probabilities:

$\frac{1}{10} \times \frac{1}{2}$

Solve the number sentences and decide which answer is reasonable.

$\frac{1}{10} \times \frac{1}{2} = \frac{1}{20}$

The probability that the spinner lands on five the first time and a number greater than five the second time is $\frac{1}{20}$.

19. *Read the entire word problem.*

We are told that a number cube, which has six sides numbered one through six, respectively, is rolled three times.

Identify the question being asked.

We are looking for the chances that it will land on an even number the first time, an odd number the second time, and the number six the third time.

Underline the keywords and words that indicate formulas.

The word *chances* tells us that we are looking for a probability, and the keyword *and* tells us that we are looking for multiple probabilities, which we will likely have to multiply.

Cross out extra information and translate words into numbers.

There is no extra information in this problem.

List the possible operations.

The total number of possibilities for each roll is the total number of sides of the cube, which is six. Since three sides contain even numbers (two, four, six), the probability of the cube landing on an even number is $\frac{3}{6}$, or $\frac{1}{2}$. Since three sides contain odd numbers (one, three, five), the probability of the cube landing on an odd number is $\frac{3}{6}$, or $\frac{1}{2}$. Only one side contains the number six, so the probability of the cube landing on the number six is $\frac{1}{6}$.

Write number sentences for each operation.

The probability that the cube lands on an even number the first time, an odd number the second time, and the number six the third time is the product of each of those probabilities:

$\frac{1}{2} \times \frac{1}{2} \times \frac{1}{6}$

Solve the number sentences and decide which answer is reasonable.

$\frac{1}{2} \times \frac{1}{2} \times \frac{1}{6} = \frac{1}{24}$

The probability that the cube lands on an even number the first time, an odd number the second time, and the number six the third time is $\frac{1}{24}$.

20. *Read the entire word problem.*

We are told that a package contains 12 chocolate cookies and ten vanilla cookies and that two cookies are taken from the package and not returned.

Identify the question being asked.

We are looking for the probability that both cookies were vanilla cookies.

Underline the keywords and words that indicate formulas.

The word *probability* tells us that we are looking for a probability, and the keyword *and* tells us that we are looking for two probabilities, which we will likely have to multiply. Since the cookies were eaten, they were not returned to the package, which tells us that the denominator of our second probability will not be the same as the denominator of our first probability, and the numerator may change as well.

Cross out extra information and translate words into numbers.

There is no extra information in this problem.

List the possible operations.

The total number of possibilities is the total number of cookies in the package at any time. To start, there are $12 + 10 = 22$ cookies in the package. Since ten of them are vanilla, the probability of Rita selecting a vanilla cookie is $\frac{10}{22}$, or $\frac{5}{11}$. Rita eats the cookie, which means that there is one less vanilla cookie (and one less total cookie), so the probability that she selects a vanilla cookie the second time is only $\frac{9}{21}$, or $\frac{3}{7}$.

Write number sentences for each operation.

The probability that Rita selects two vanilla cookies is the product of each of those probabilities:

$\frac{5}{11} \times \frac{3}{7}$

Solve the number sentences and decide which answer is reasonable.

$\frac{5}{11} \times \frac{3}{7} = \frac{15}{77}$

The probability that Rita selects two vanilla cookies is $\frac{15}{77}$.

21. *Read the entire word problem.*

We are told that a program offers five science courses, eight math courses, four art courses, and three music courses, and that Mairead is placed in two courses that are not identical.

Identify the question being asked.

We are looking for the odds that she has been placed in an art course and a music course.

Underline the keywords and words that indicate formulas.

The word *odds* tells us that we are looking for a probability, and the keyword *and* tells us that we are looking for two probabilities, which we will likely have to multiply. Since she cannot be placed in the same course twice, the denominator of our second probability will not be the same as the denominator of our first probability.

Cross out extra information and translate words into numbers.

There is no extra information in this problem.

List the possible operations.

The total number of possibilities is the total number of courses at any time. To start, there are $5 + 8 + 4 + 3 = 20$ courses. Since four of them are art courses, the probability of Mairead being placed in an art course is $\frac{4}{20}$, or $\frac{1}{5}$. After Mairead is placed in that art course, she cannot be placed in that course again, so there are now only 19 courses left. The probability that she is placed in a music course is $\frac{3}{19}$, since there are three music courses offered out of the remaining 19 courses.

Write number sentences for each operation.

The probability that Mairead has been placed in an art course and a music course is the product of those probabilities:

$\frac{1}{5} \times \frac{3}{19}$

Solve the number sentences and decide which answer is reasonable.

$\frac{1}{5} \times \frac{3}{19} = \frac{3}{95}$

The probability that Mairead is placed in an art course and a music course is $\frac{3}{95}$.

22. *Read the entire word problem.*

We are told that a sock drawer contains eight pairs of brown socks, 18 pairs of white socks, and 20 pairs of black socks and that each of two girls selects a pair of socks.

Identify the question being asked.

We are looking for the likelihood that Molly will select a white pair and that Anna will select a black pair.

Underline the keywords and words that indicate formulas.

The word *likelihood* tells us that we are looking for a probability, and the keyword *and* tells us that we are looking for two probabilities, which we will likely have to multiply. Since Molly and Anna cannot both wear the same pair of socks, the denominator of our second probability will not be the same as the denominator of our first probability.

Cross out extra information and translate words into numbers.

There is no extra information in this problem.

List the possible operations.

The total number of possibilities is the total number of socks from which to choose at any time. To start, there are $8 + 18 + 20 = 46$ pairs of socks. Since 18 of them are white, the probability of Molly selecting a white pair is $\frac{18}{46}$, or $\frac{9}{23}$. After Molly selects a pair of white socks, there is one fewer pair of socks in the drawer. Anna has 45 pairs of socks from which to choose, and 20 of them are black, so the probability that she selects a black pair of socks is $\frac{20}{45}$, or $\frac{4}{9}$.

Write number sentences for each operation.

The probability that Molly will select a white pair and that Anna will select a black pair is the product of those probabilities:

$\frac{9}{23} \times \frac{4}{9}$

Solve the number sentences and decide which answer is reasonable.

$\frac{9}{23} \times \frac{4}{9} = \frac{36}{207} = \frac{4}{23}$

The probability that Molly will select a white pair and that Anna will select a black pair is $\frac{4}{23}$.

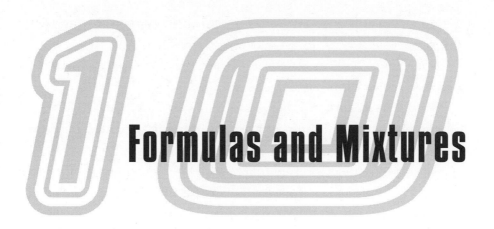

Formulas and Mixtures

Some word problems can be solved using a specific formula. We must use the context and keywords of the problem to figure out which formula to use, and then substitute the values given in the problem into the formula to solve for the missing value. We'll look at some simple formulas, and then a more complex type of a word problem—a mixture problem.

FORMULAS: SUBJECT REVIEW

The formula for distance is $D = rt$, where D, distance, is equal to the product of r, rate, and t, time.

If an object travels at 30 miles per hour for three hours, it will travel 30 × 3 = 90 miles.

Sometimes we need to rearrange a formula to solve for a certain value. For instance, $D = rt$, but if we are given distance and rate and are looking for time, we must rewrite the formula to solve for time. Dividing both sides of the equation by r gives us $t = \frac{D}{r}$. To find time given distance and rate, we must divide the distance by the rate.

In the same way, $I = prt$, but if we are looking for the principal given the interest, rate, and time, we must divide both sides of the formula by r and t to get p alone on one side of the equation. Principal is equal to $\frac{I}{rt}$. The formula for interest is $I = prt$, where I, interest, is equal to the product of p, principal,

r, rate, and t, time. The rate is given as a percent and should be converted to a decimal.

A principal of $500, at a rate of 4% per year for six years, will make ($500)(.04)(6) = $120.

Given a temperature in Fahrenheit, F, we can convert to Celsius, C, or given a temperature in Celsius, we can convert to Fahrenheit: $C = \frac{5}{9}(F - 32)$ and $F = \frac{9}{5}C + 32$.

If the temperature is 86° Fahrenheit, then the temperature in Celsius is $(86 - 32) = \frac{5}{9}(54) = 30°$. If the temperature is 20° Celsius, then the temperature in Fahrenheit is $\frac{9}{5}(20) + 32 = 36 + 32 = 68°$.

The formula for velocity is $V = at$, where V, velocity, is equal to the product of a, acceleration, and t, time.

If an object at rest accelerates at 5 meters per second for ten seconds, then its velocity is $(5)(10) = 50$ meters per second.

There are many formulas for converting one unit of measure to another. Here are just a few:

- To convert from grams, meters, or liters to milligrams, millimeters, or milliliters, multiply by 1,000.
- To convert from milligrams, millimeters, or milliliters to grams, meters, or liters divide by 1,000.
- To convert from inches to feet, divide the number of inches by 12.
- To convert from feet to inches, multiply the number of feet by 12.
- To convert from feet to yards, divide the number of feet by 3.
- To convert from yards to feet, multiply the number of yards by 3.
- To convert from ounces to pounds, divide the number of ounces by 16.
- To convert from pounds to ounces, multiply the number of pounds by 16.

Example

David's model car travels at 2 feet per second. If the car travels for ten seconds, how far will it travel?

Read the entire word problem.
We are given the rate at which a model car travels and the time in which it travels.
Identify the question being asked.

We are looking for the distance it travels.

Underline the keywords.

The keyword *per* can signal multiplication or division.

Cross out extra information and translate words into numbers.

There is no extra information in this problem.

List the possible operations.

To find the distance an object travels, we must multiply the rate by the time. The keyword *per* signals multiplication, since the rate gives us the distance traveled in one second, and we are looking for the distance traveled in ten seconds.

Write number sentences for each operation.

Our number sentence is the distance formula, with 2 substituted for *r* and 10 substituted for *t*: $D = (2)(10)$

Solve the number sentences and decide which answer is reasonable.

20 feet = (2)(10)

Check your work.

Since $D = rt$, we can check our distance answer by dividing the distance by the rate to see if that quotient equals the given time, 10 seconds: $\frac{20}{2} = 10$ seconds. We also could have divided distance by the time to see if that quotient equals the given rate, 2 feet per second: $\frac{20}{10}$ = 2 feet per second.

At the same time we study a word problem for keywords, we should also search for words that tell us to use a specific formula. The word *per* can mean "rate," and the word *travel* often means that we are working with a distance. The words *Celsius* and *Fahrenheit* may indicate that we need to convert from one unit to the other.

Example

Kimberly hears that the temperature in Sydney, Australia, is 22° Celsius. What is the temperature in Sydney in degrees Fahrenheit?

Read the entire word problem.

We are given a temperature in degrees Celsius.

Identify the question being asked.

We are looking for a temperature in degrees Fahrenheit.

Underline the keywords and words that indicate formulas.

The words *Celsius* and *Fahrenheit* tell us that we need to use a temperature conversion formula.

Cross out extra information and translate words into numbers.

There is no extra information in this problem.

List the possible operations.

We will use multiplication and addition to convert from degrees Celsius to degrees Fahrenheit.

Write number sentences for each operation.

Substitute the temperature in degrees Celsius into the formula:

$F = \frac{9}{5}(22) + 32$

Solve the number sentences and decide which answer is reasonable.

$F = \frac{9}{5}(22) + 32 = 39.6 + 32 = 71.6°$

Check your work.

We can use the formula for converting Fahrenheit to Celsius to check our answer. $C = \frac{5}{9}(F - 32) = \frac{5}{9}(71.6 - 32) = \frac{5}{9}(39.6) = 22°$ Celsius

---

PRACTICE LAP

DIRECTIONS: Solve the following word problems using the eight-step process.

1. Bettyann puts $2,000 in the bank and gains interest at a rate of 4.5% per year. How much interest has she earned after five years?

2. Carlos throws a paper airplane across a room. If the plane accelerates from rest at a rate of 3.2 feet per second squared, what is the velocity of the plane after three seconds?

3. Dimitri is riding a train that is traveling at 82 miles per hour. If he rides the train for two and a quarter hours, how far will he have traveled?

4. Shelly's thermometer shows that it is 65° Fahrenheit today. If it was 5° Fahrenheit cooler yesterday, what was the temperature in degrees Celsius yesterday?

5. If a jug weighs 86.4 ounces, how many pounds does it weigh?

Example

Dianne's pencil is 176 millimeters long. How long is her pencil in meters?

Let's say we need to work quickly and do not have time for the eight-step process. This problem is asking us to convert millimeters to meters. To convert from millimeters to meters, we must divide the number of millimeters by 1,000: $\frac{176}{1,000} = 0.176$ meters. Dianne's pencil is 0.176 meters long.

Example

Dianne earns $228 in interest on $1,500 that she kept in a bank for four years. What was the interest rate on Dianne's principal?

We know that $I = prt$, so to solve for r, the interest rate, we must rearrange the equation. If we divide both sides of the equation by p and t, we have $r = \frac{I}{pt}$. Substitute the interest, principal, and time into the formula to find the interest rate: $\frac{228}{(1,500)(4)} = \frac{228}{6,000} = 0.038 = 3.8\%$.

CAUTION

READ THE UNITS of each value in a word problem carefully. If a distance problem gives you a rate in meters per second and a time in seconds, you are ready to multiply, but if the rate is given in meters per second and the time is given in minutes, you must either convert the rate to meters per minutes or convert the time to seconds before multiplying. In the same way, if you are looking to find interest earned, the rate and time must also be given in the same units. You may need to convert one number from months to years.

Example

Fernando throws a baseball at a speed of 90 miles per hour. How many seconds does it take the ball to travel 60.5 feet?

We are given a rate and a distance, and we are looking for a time. However, our rate is given in miles per hour and our distance is given in feet. We must either convert our rate to feet per hour or convert the distance to miles. Since there are 5,280 feet in a mile, 60.5 feet is equal to $\frac{60.5}{5,280}$, or approximately 0.0116 miles. The distance formula is $D = rt$, but we are looking for t, the time. Divide both sides of the equation by r, and $t = \frac{D}{r} = \frac{0.0116}{90}$, which equals 0.000129. Our rate was given in miles per hour, so this is the number of hours it takes the ball to travel 60.5 feet. There are 60 minutes in an hour and 60 seconds in a minute, which means that we must multiply the time in hours by $60 \times 60 = 3,600$. So, $0.000129 \times 3,600 = 0.4644$ seconds.

PRACTICE LAP

DIRECTIONS: Try to solve these word problems by identifying which formula to use. If you have trouble, use the eight-step process.

6. How many feet are in 40 yards?

7. A rock falling from a cliff accelerates from rest at a rate of 9.8 meters per second squared. How many seconds will it take for the rock to reach a velocity of 31.36 meters per second?

8. If a plane travels 455 miles per hour, how long will it take to travel 2,375 miles? Round your answer to the nearest hundredth of an hour.

9. Juan puts $5,000 in a bank account that gains interest at a rate of 4.7% per year. How much interest will Juan have earned after six months?

10. One lap at the Atlanta Motor Speedway is 1.54 miles. If Travis completes a lap in 30.8 seconds, how fast, in miles per hour, was he traveling?

PACE YOURSELF

USE A MAP to find how far you live from your school. Use the time it takes you to walk to school to find your average rate of speed, in miles per hour. If you take a car or bus to school, use that time to find the average rate of speed in miles per hour of the car or bus.

MIXTURE PROBLEMS

A mixture problem is just like what it sounds—when two quantities are mixed together. For example, a salt solution is a mixture of salt and water. We can change the solution by adding or removing salt or water. Or we could mix a solution that has a lot of salt in it with a solution that has only a little salt in it to make a new solution.

We solve mixture problems by making tables to organize the data. Let's look at an example.

Example

Miguel has 10 ounces of a solution of salt and water that is 30% salt. He has 10 ounces of this solution, but he wants the solution to be 35% salt. How much salt must he add to the solution?

Our table will have three rows and three columns. We will have a row for our original solution, a row for what we are adding to the solution, and a row for our new, final solution. We will have a column to hold the percent concentration of salt in each solution, a column to hold the total quantity of each solution, and a column to hold the salt quantity of each solution:

	Percent concentration	Total quantity (oz.)	Salt quantity (oz.)
Original solution			
Added			
Final solution			

The original solution is 10 ounces and 30% salt. Enter these figures in the table after converting 30% to a decimal. The number of ounces of salt in the solution is equal to the total number of ounces multiplied by the percent of ounces that are salt (10×0.30):

	Percent concentration	Total quantity (oz.)	Salt quantity (oz.)
Original solution	0.30	10	0.3×10
Added			
Final solution			

To this solution, we must add salt to make a solution that is 35% salt. We don't know how many ounces of salt to add, so we represent that quantity with x. The percent concentration of salt that we are adding is 100%, since we are adding pure salt to the solution. The salt quantity of what we are adding is equal to 100% of x, or $1 \times x$:

	Percent concentration	Total quantity (oz.)	Salt quantity (oz.)
Original solution	0.30	10	0.3×10
Added	1.00	x	$1 \times x$
Final solution	0.35		

The total quantity in ounces of the final solution is equal to the total quantity in ounces of the original solution plus the quantity of salt in ounces that was added: $10 + x$. The salt quantity of the final solution is equal to 0.35 multiplied by the total quantity, $10 + x$:

	Percent concentration	Total quantity (oz.)	Salt quantity (oz.)
Original solution	0.30	10	0.3×10
Added	1.00	x	$1 \times x$
Final solution	0.35	$10 + x$	$0.35 \times (10 + x)$

The salt quantity of the final solution is also equal to the salt quantity of the original solution plus the salt quantity that was added: $(0.3 \times 10) + (1 \times x) = 3 + x$. Since the salt quantity of the final solution is equal to $3 + x$ and $0.35(10 + x)$, we can set these two values equal to each other and solve for x, the number of ounces of salt added:

$3 + x = 0.35(10 + x)$

$3 + x = 3.5 + 0.35x$

$3 - 3 + x = 3.5 - 3 + 0.35x$

$x = 0.5 + 0.35x$

$0.65x = 0.5$

$x = \frac{50}{65} = \frac{10}{13}$

Miguel should add $\frac{10}{13}$ ounces of salt to the solution.

We can also remove water from a solution to change the concentration of the solution.

Example

Eli has 50 ounces of an alcohol solution that is 20% alcohol. How many ounces of water must be evaporated to make the concentration 40% alcohol?

Our table again will have three rows and three columns. Since we are looking for how much water will be evaporated, our table will be about the percent and quantity of water. Since the solution is 20% alcohol, it is 100% − 20% = 80% water, and we are looking for a final solution that is only 100% − 40% = 60% water:

	Percent concentration	Total quantity (oz.)	Water quantity (oz.)
Original solution	0.80	50	0.8 × 50
Removed	1	x	1 × x
Final solution	0.60		

Remember, we are removing water, so the total quantity will be 50 − x, not 50 + x:

	Percent concentration	Total quantity (oz.)	Water quantity (oz.)
Original solution	0.80	50	0.8 × 50
Removed	1	x	1 × x
Final solution	0.60	50 − x	0.60(50 − x)

The final water quantity is equal to $0.60(50 - x)$ and $(0.8 \times 50) - (1 \times x)$, which is the water quantity of the original solution minus the quantity that was removed (evaporated). We can set these equations equal to each other and solve for x:

$$0.60(50 - x) = (0.8 \times 50) - (1 \times x)$$
$$30 - 0.6x = 40 - x$$
$$0.4x = 10$$
$$x = 25$$

To make the concentration of the solution 40% alcohol, 25 ounces of water must be evaporated.

Let's look at one last type of mixture problem: mixing two different solutions together.

Example

Dawn has 15 liters of a 20% alcohol solution. How many liters of 50% alcohol solution must be added to create a 30% alcohol solution?

We have 15 liters of a 0.20 solution, which means that the alcohol quantity of that solution is 0.2×15. We are adding an unknown quantity of a 0.50 solution, which has an alcohol quantity of $0.5 \times x$, and our final solution will be a 0.30 alcohol solution, with an alcohol quantity of 0.30 times the total quantity, $15 + x$:

	Percent concentration	Total quantity (L)	Alcohol quantity (L)
Original (20%) solution	0.20	15	0.2×15
Added (50%) solution	0.50	x	$0.5 \times x$
Final (30%) solution	0.30	$15 + x$	$0.3(15 + x)$

The variable x represents how many liters of the 50% solution we must add. Just as in the last examples, look at the final column. The sum of the first two rows, 0.2×15 and $0.5 \times x$, is equal to the product of the two columns of the last row, 0.30 and $15 + x$, since both expressions are equal to the alcohol quantity of the final solution:

$$(0.2 \times 15) + (0.5 \times x) = 0.3(15 + x)$$
$$3 + 0.5x = 4.5 + 0.3x$$
$$0.2x = 1.5$$
$$x = 7.5$$

Dawn must add 7.5 liters of 50% alcohol solution to the 20% alcohol solution to create a 30% alcohol solution.

- PRACTICE LAP - - - - - - - - -

DIRECTIONS: Round your answers to the nearest hundredth.

11. Raymar's 25-ounce salt solution is 25% salt. How much salt must he add to the solution to make it 30% salt?

12. Rosaria has a 40-ounce salt solution that is 10% salt. How much salt must she add to the solution to make it 25% salt?

13. How many liters of water must be evaporated from 30 liters of a 45% alcohol solution to make the concentration 55% alcohol?

14. Yao has 5 liters of a solution that is 55% alcohol. How many liters of water must be evaporated to make the concentration 80% alcohol?

15. Libby has 22 liters of a 40% alcohol solution. How many liters of 70% alcohol solution must be added to create a 60% alcohol solution?

16. Jana has 16 liters of a 5% alcohol solution. How many liters of a 45% alcohol solution must be added to create an 18% alcohol solution?

PACE YOURSELF

TRY MAKING A mixture yourself. Fill a beaker with 8 ounces of water, and add 2 ounces of salt. The beaker is now a 20% salt solution. How much salt must you add to make a 50% salt solution?

SUMMARY

WE CAN SOLVE some word problems using formulas. After reviewing some common formulas, we made a small change to our eight-step process to look for words that signal which of these formulas to use. To save time, we learned to substitute values into the formula rather than using the entire eight-step process. We also learned how to rearrange formulas to solve for each part of a formula. Mixture problems are easier to solve with a table than with the eight-step process, so we learned how to set up a specific kind of table to solve this type of problem.

ANSWERS

1. *Read the entire word problem.*

   We are given the principal, rate, and time.

   Identify the question being asked.

   We are looking for the interest earned by that principal at that rate over that time.

   Underline the keywords and words that indicate formulas.

   The keyword *per* can signal multiplication or division. The words *interest* and *rate* tell us that we need to use the formula for interest.

   Cross out extra information and translate words into numbers.

   There is no extra information in this problem.

   List the possible operations.

   To find interest earned given the principal, rate, and time, we must multiply those three values.

   Write number sentences for each operation.

   Convert 4.5% to a decimal. Our number sentence is the interest formula, with 2,000 substituted for p, 0.045 substituted for r, and 5 substituted for t: $I = (2,000)(0.045)(5)$.

   Solve the number sentences and decide which answer is reasonable.

   $I = (2,000)(0.045)(5) = \$450$

   Check your work.

   Since $I = prt$, we can check our interest answer by dividing the interest by the rate and time to see if that quotient equals the given principal, \$2,000: $\frac{450}{(.045)(5)} = \frac{450}{0.225} = \$2,000$.

2. *Read the entire word problem.*

   We are given the acceleration and hang time of a paper airplane.

   Identify the question being asked.

We are looking for the velocity of the plane after three seconds.

Underline the keywords and words that indicate formulas.

The keyword *per* can signal multiplication or division. The words *accelerates* and *rate* tell us that we need to use the velocity formula.

Cross out extra information and translate words into numbers.

There is no extra information in this problem.

List the possible operations.

To find velocity given the acceleration and time, we must multiply the acceleration by the time.

Write number sentences for each operation.

Our number sentence is the velocity formula, with 3.2 substituted for *a* and 3 substituted for *t*: $V = (3.2)(3)$.

Solve the number sentences and decide which answer is reasonable.

$V = (3.2)(3) = 9.6$ feet per second

Check your work.

Since $V = at$, we can check our velocity answer by dividing the velocity by the acceleration to see if that quotient equals the given time, 3 seconds: $\frac{9.6}{3.2} = 3$ seconds.

3. *Read the entire word problem.*

We are given the rate at which a train is traveling and the time the train travels at that rate.

Identify the question being asked.

We are looking for the distance traveled by Dimitri.

Underline the keywords and words that indicate formulas.

The keyword *per* can signal multiplication or division. The word *travel* could mean that we need to use the distance formula.

Cross out extra information and translate words into numbers.

There is no extra information in this problem, but we must translate *two and a quarter* into 2.25.

List the possible operations.

To find distance given the rate and time, we must multiply the rate by the time.

Write number sentences for each operation.

Our number sentence is the distance formula, with 82 substituted for *r* and 2.25 substituted for *t*: $D = (82)(2.25)$.

Solve the number sentences and decide which answer is reasonable.

$D = (82)(2.25) = 184.5$ miles

Check your work.

Since $D = rt$, we can check our distance answer by dividing the distance by the rate to see if that quotient equals the given time, 2.25 hours: $\frac{184.5}{82}$ = 2.25 hours.

4. *Read the entire word problem.*

We are given a temperature in degrees Fahrenheit and the number of degrees the temperature increased from the previous day.

Identify the question being asked.

We are looking for yesterday's temperature in degrees Celsius.

Underline the keywords and words that indicate formulas.

The words *Celsius* and *Fahrenheit* tell us that we need to use a temperature conversion formula.

Cross out extra information and translate words into numbers.

There is no extra information in this problem.

List the possible operations.

The temperature was 5° cooler the previous day, so we must first find the temperature in degrees Fahrenheit yesterday, and then convert that to degrees Celsius.

Write number sentences for each operation.

The temperature was 5° cooler yesterday, so we must subtract 5 from 65 to find the temperature yesterday:

$65 - 5$

Solve the number sentences and decide which answer is reasonable.

$65 - 5 = 60°$

Write number sentences for each operation.

Now that we have the temperature yesterday in degrees Fahrenheit, we can convert it to degrees Celsius. Substitute the temperature in degrees Fahrenheit into the formula:

$C = \frac{5}{9}(60 - 32)$

Solve the number sentences and decide which answer is reasonable.

$C = \frac{5}{9}(60 - 32) = \frac{5}{9}(28) \times 15.56°$

Check your work.

We can use the formula for converting Celsius to Fahrenheit to check our answer. $F = \frac{9}{5}C + 32$, $F = \frac{9}{5}(15.56) + 32 = 28 + 32 = 60°$. Since the temperature that day was 5° cooler than today, add 5 to 60. This should equal the temperature in degrees Fahrenheit today, 65: 5 + 60 = 65° Fahrenheit.

5. *Read the entire word problem.*
   We are given a weight in ounces.
   Identify the question being asked.
   We are looking for that weight in pounds.
   Underline the keywords and words that indicate formulas.
   The words *ounces* and *pounds* tell us that we need to use a weight conversion formula.
   Cross out extra information and translate words into numbers.
   There is no extra information in this problem.
   List the possible operations.
   Since 1 ounce weighs less than 1 pound, we must divide the number of ounces to find the number of pounds.
   Write number sentences for each operation.
   There are 16 ounces in 1 pound, so we must divide the number of ounces by 16 to find the number of pounds:
   $\frac{86.4}{16}$
   Solve the number sentences and decide which answer is reasonable.
   $\frac{86.4}{16} = 5.4$ pounds
   Check your work.
   We can check our work by converting the number of pounds back to ounces. Multiply the number of pounds by 16, and that should equal the weight of the jug in ounces, 86.4; 5.4 × 16 = 86.4 ounces.

6. There are 3 feet in 1 yard, so we must multiply the number of yards by 3 to find the number of feet: (40)(3) = 120 feet.

7. We are given an acceleration and a velocity, so we must use the velocity formula to find the time. Since $V = at$, we must divide both sides of the equation by a to find the formula for t, time: $t = \frac{V}{a} = \frac{31.36}{9.8} = 3.2$ seconds.

8. We are given a rate of speed and a distance, so we must use the distance formula to find the time. Since $D = rt$, we must divide both sides of the equation by r to find the formula for t, time: $t = \frac{D}{r} = \frac{2,375}{455} \times 5.22$ hours.

9. We are given a principal, a rate, and a time, so we must use the interest formula to find the interest. However, we are given the rate in years and the time in months, so we must convert the time to years. There are

12 months in a year, so six months is $\frac{6}{12} = 0.5$ years. Now we are ready to use the formula: $I = prt = (5,000)(0.047)(0.5) = \$117.50$.

10. We are given a distance and a time, so we must use the distance formula to find the rate of speed. Since $D = rt$, we must divide both sides of the equation by t to find the formula for r, rate: $r = \frac{D}{t} = \frac{1.54}{30.8} = 0.05$. However, since we were given the time in seconds, this rate is the number of miles per second, not miles per hour. There are 60 seconds in a minute and 60 minutes in an hour, so to convert miles per second to miles per hour, we must multiply 0.05 by 3,600: $0.05 \times 3,600 = 180$ miles per hour.

11. Make a table with three rows and three columns. Since we are looking for how much salt will be added, our table will be about the percent and quantity of salt. The original solution is 25%, or 0.25, salt. It is 25 ounces in total, which means that 0.25×25 of it is salt. We are adding an unknown quantity of salt, 100% of which is salt. Our final solution will be 30% salt:

	Percent concentration	Total quantity (oz.)	Salt quantity (oz.)
Original solution	0.25	25	0.25×25
Added	1	x	$1 \times x$
Final solution	0.30		

The total quantity of the final solution will be $25 + x$, and 0.30 of it will be salt:

	Percent concentration	Total quantity (oz.)	Salt quantity (oz.)
Original solution	0.25	25	0.25×25
Added	1	x	$1 \times x$
Final solution	0.30	$25 + x$	$0.30(25 + x)$

The salt quantity equals $0.30(25 + x)$ and also equals the sum of 0.25×25 and $1 \times x$. Set these two expressions equal to each other and solve for x:

$0.30(25 + x) = (0.25 \times 25) + (1 \times x)$

$7.5 + 0.3x = 6.25 + x$

$1.25 = 0.7x$

$x = 1.79$ ounces

12. Make a table with three rows and three columns. Since we are looking for how much salt will be added, our table will be about the percent and quantity of salt. The original solution is 10%, or 0.10, salt. It is 40 ounces in total, which means that 0.10×40 of it is salt. We are adding an unknown quantity of salt, 100% of which is salt. Our final solution will be 25% salt:

	Percent concentration	Total quantity (oz.)	Salt quantity (oz.)
Original solution	0.10	40	0.10×40
Added	1	x	$1 \times x$
Final solution	0.25	x	

The total quantity of the final solution will be $40 + x$, and 0.25 of it will be salt:

	Percent concentration	Total quantity (oz.)	Salt quantity (oz.)
Original solution	0.10	40	0.10×40
Added	1	x	$1 \times x$
Final solution	0.25	$40 + x$	$0.25(40 + x)$

The salt quantity is the product of the percent concentration and the total quantity: $0.25(40 + x)$. It is also the sum of the salt quantity in the original solution and the amount added: 0.10×40 and $1 \times x$. Set these two expressions equal to each other and solve for x:

$0.25(40 + x) = (0.10 \times 40) + (1 \times x)$

$10 + 0.25x = 4 + x$

$6 = 0.75x$

$x = 8$ ounces

13. Make a table with three rows and three columns. Since we are looking for how much water will be evaporated, our table will be about the percent and quantity of water. The solution is 45% alcohol and $100\% - 45\% = 55\%$ water. We are looking for a final solution that is only $100\% - 55\% = 45\%$ water:

	Percent concentration	Total quantity (L)	Water quantity (L)
Original solution	0.55	30	0.55×30
Removed	1	x	$1 \times x$
Final solution	0.45		

Remember, we are removing water, so the total quantity will be $30 - x$, not $30 + x$:

	Percent concentration	Total quantity (L)	Water quantity (L)
Original solution	0.55	30	0.55×30
Removed	1	x	$1 \times x$
Final solution	0.45	$30 - x$	$0.45(30 - x)$

The final water quantity is equal to the percent concentration of the final solution times the total quantity of the solution: $0.45(30 - x)$. The final quantity of water is also equal to the original quantity of water minus the amount of water removed: $(0.55 \times 30) - (1 \times x)$. Set these equations equal to each other and solve for x:

$0.45(30 - x) = (0.55 \times 30) - (1 \times x)$

$13.5 - 0.45x = 16.5 - x$

$0.55x = 3$

$x = 5.45$ liters, to the nearest hundredth

14. Make a table with three rows and three columns. The solution is 55% alcohol and $100\% - 55\% = 45\%$ water. We are looking for a final solution that is only $100\% - 80\% = 20\%$ water:

	Percent concentration	Total quantity (L)	Water quantity (L)
Original solution	0.45	5	0.45×5
Removed	1	x	$1 \times x$
Final solution	0.20		

Remember, we are removing water, so the total quantity will be $5 - x$, not $5 + x$:

	Percent concentration	Total quantity (L)	Water quantity (L)
Original solution	0.45	5	0.45×5
Removed	1	x	$1 \times x$
Final solution	0.20	$5 - x$	$0.20(5 - x)$

The final water quantity is equal to the percent concentration of the final solution times the total quantity of the solution: $0.20(5 - x)$. The final quantity of water is also equal to the original quantity of water minus the amount of water removed: $(0.45 \times 5) - (1 \times x)$. Set these equations equal to each other and solve for x:

$0.20(5 - x) = (0.45 \times 5) - (1 \times x)$

$1 - 0.20x = 2.25 - x$

$0.8x = 1.25$

$x = 1.56$ liters, to the nearest hundredth.

15. Libby has 22 liters of a 0.40 solution, which means that the alcohol quantity of that solution is 0.4×22. We are adding an unknown quantity of a 0.70 solution, which has an alcohol quantity of $0.7 \times x$, and our final solution will be a 0.60 alcohol solution, with an alcohol quantity of 0.60 times the total quantity, $22 + x$:

	Percent concentration	Total quantity (L)	Alcohol quantity (L)
Original (40%) solution	0.40	22	0.4×22
Added (70%) solution	0.70	x	$0.7 \times x$
Final (60%) solution	0.60	$22 + x$	$0.6(22 + x)$

The variable x represents how many liters of the 70% solution we must add. The final alcohol quantity is equal to the alcohol quantity of the original solution plus the alcohol quantity of the added solution: 0.4×22 and $0.7 \times x$. The final alcohol quantity is also equal to the product of percent concentration of the final solution and the total quantity of that solution: $0.6(22 + x)$. Set these equations equal to each other and solve for x:

$(0.4 \times 22) + (0.7 \times x) = 0.6(22 + x)$

$8.8 + 0.7x = 13.2 + 0.6x$

$0.1x = 4.4$

$x = 44$ liters

Libby must add 44 liters of 70% alcohol solution to the 40% alcohol solution to create a 60% alcohol solution.

16. Jana has 16 liters of a 0.05 solution, which means that the alcohol quantity of that solution is 0.05×16. We are adding an unknown quantity of a 0.45 solution, which has an alcohol quantity of $0.45 \times x$, and our final solution will be an 0.18 alcohol solution, with an alcohol quantity of 0.18 times the total quantity, $16 + x$:

	Percent concentration	Total quantity (L)	Alcohol quantity (L)
Original (5%) solution	0.05	16	0.05×16
Added (45%) solution	0.45	x	$0.45 \times x$
Final (18%) solution	0.18	$16 + x$	$0.18(16 + x)$

The variable x represents how many liters of the 45% solution Jana must add. The final alcohol quantity is equal to the alcohol quantity of the original solution plus the alcohol quantity of the added solution: 0.05×16 and $0.45 \times x$. The final alcohol quantity is also equal to the product of percent concentration of the final solution and the total quantity of that solution: $0.18(16 + x)$. Set these equations equal to each other and solve for x:

$(0.05 \times 16) + (0.45 \times x) = 0.18(16 + x)$

$0.8 + 0.45x = 2.88 + 0.18x$

$0.27x = 2.08$

$x = 7.70$ liters, to the nearest hundredth

Jana must add approximately 7.70 liters of 45% alcohol solution to the 5% alcohol solution to create an 18% alcohol solution.

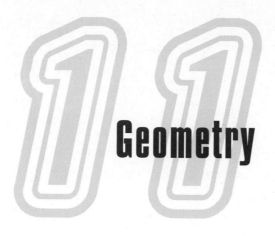

Geometry

While probability problems are usually word problems, geometry problems usually have a diagram. They can be a bit tougher when there are no pictures, just words, to describe a shape and its dimensions. We can use the eight-step process to solve a geometry problem, but often it's better to draw a picture. Sometimes, we may even want to do both.

ANGLES: SUBJECT REVIEW

When two parallel lines are cut by a transversal, **adjacent angles**, **corresponding angles**, **vertical angles**, and **supplementary angles** are created. Corresponding angles, or alternating angles, are **congruent** to each other, as are vertical angles.

If two angles, lines, or shapes are equal in measure, then they are **congruent**.

Corresponding angles are two or more congruent angles that have similar positions in a shape or figure.

Vertical angles are congruent angles that are formed by the intersection of two lines. Vertical angles are opposite to each other, which is why they are sometimes called **opposite angles**.

Supplementary angles are two angles whose measures total 180°.

A diagram or shape can also contain **acute**, **obtuse**, or **right** angles. Two right angles are **complementary** to each other.

FUEL FOR THOUGHT

A TRIANGLE IS a three-sided polygon whose angles total 180°.

An **acute angle** measures less than 90°. At least two angles in a triangle are acute. If all three angles are less than 90°, then the triangle is an **acute triangle**.

A **right angle** measures exactly 90°. No more than one angle in a triangle can be a right angle. If a triangle does contain a 90° angle, then it is a **right triangle**.

An **obtuse angle** measures greater than 90°. No more than one angle in a triangle can be an obtuse angle. If a triangle does contain an angle that is greater than 90°, then it is an **obtuse triangle**.

Complementary angles are two angles whose measures total 90°.

An **equilateral triangle** has three congruent sides and three congruent, 60° angles.

An **isosceles triangle** has two congruent sides and two congruent angles.

When we are working on an angle word problem where one single angle measure is given and we are looking for the measure of another angle, it's likely our answer will be either (1) the same measure as the angle given, (2) the complement of the given angle, or (3) the supplement of the given angle. The keywords *corresponding*, *alternating*, and *vertical* can signal that two angles are congruent. The keyword *supplementary* can mean that we must subtract the measure of one angle from 180°, and the keyword *complementary* can indicate that we must subtract the measure of one angle from 90°.

Example

> If angles *A* and *B* are vertical angles, and the measure of angle *A* is 65°, what is the measure of angle *B*?

> *Read the entire word problem.*
> We are given the measure of angle *A*, and we are told that it and angle *B* are vertical angles.
> *Identify the question being asked.*
> We are looking for the measure of angle *B*.
> *Underline the keywords and words that indicate formulas.*
> The word *vertical* tells us that angles *A* and *B* are congruent—their measures are the same.
> *Cross out extra information and translate words into numbers.*
> There is no extra information in this problem.
> *List the possible operations.*
> We don't need to perform any operations or write any number sentences. Vertical angles are congruent, so since angle *A* is 65°, angle *B* is 65°.
> *Check your work.*
> We didn't perform any operations, so we don't have any work to check. We can check the definition of vertical angles, which states that two vertical angles are equal in measure.

Example

> If angles *A* and *B* are complementary angles, and the measure of angle *A* is 12°, what is the measure of angle *B*?

Read the entire word problem.

We are given the measure of angle *A*, and we are told that it and angle *B* are complementary angles.

Identify the question being asked.

We are looking for the measure of angle *B*.

Underline the keywords and words that indicate formulas.

The word *complementary* tells us that the measures of angles *A* and *B* total 90°.

Cross out extra information and translate words into numbers.

There is no extra information in this problem.

List the possible operations.

Since complementary angles add to 90° and we have the measure of one angle, we must use subtraction to find the measure of the other angle.

Write number sentences for each operation.

90 – 12

Solve the number sentences and decide which answer is reasonable.

90 – 12 = 78°

Check your work.

Since complementary angles add to 90°, the sum of the measures of angles *A* and *B* should be 90°: 78 + 12 = 90°.

Not all geometry questions require computations. Sometimes, a word problem may test our knowledge of a rule or property. Look closely at the question being asked to see if it is asking you for a number, word, or phrase.

Example

If a triangle contains a 91° angle, what kind of triangle is it?

Read the entire word problem.

We are given the measure of one angle of a triangle.

Identify the question being asked.

We must identify what kind of triangle it is.

Underline the keywords and words that indicate formulas.

There are no keywords in this problem.

Cross out extra information and translate words into numbers.

There is no extra information in this problem.

List the possible operations.

A 91° angle is an obtuse angle, because it is greater than 90° in measure. This means that the triangle is an obtuse triangle. We don't need to write any number sentences.

Check your work.

The definition of an obtuse triangle is a triangle that contains an angle that is greater than 90°, so our answer is correct.

CAUTION

EVEN IF A word problem does require addition, subtraction, multiplication, or division, sometimes our answer can still be a word or phrase and not a number. The first step of the word-problem-solving process is always the most important: read the entire word problem. Equally important is determining the question being asked so that we know how to go about solving the problem.

Example

If a triangle contains two 33° angles, in what two ways can this triangle be classified?

Read the entire word problem.

We are given the measure of two angles of a triangle.

Identify the question being asked.

We are looking for two ways to name the triangle.

Underline the keywords and words that indicate formulas.

There are no keywords in this problem.

Cross out extra information and translate words into numbers.

There is no extra information in this problem.

List the possible operations.

Triangles can be classified as acute, right, obtuse, equilateral, and isosceles. A triangle is equilateral if all three angles measure 60°. Since two of the angles in this triangle are 33°, the triangle is not equilateral. However, the triangle is isosceles, because an isosceles triangle has two congruent angles. We have one name for this triangle, but we need another. In order to tell if this triangle is acute, right, or obtuse, we must find the measure of the third angle. There are 180° in a triangle, so we must find the sum of the two 33° angles, and subtract it from 180°.

Write number sentences for each operation.

33 + 33

Solve the number sentences and decide which answer is reasonable.

33 + 33 = 66°

Write number sentences for each operation.

Now subtract that sum from 180°:

180 − 66

Solve the number sentences and decide which answer is reasonable.

180 − 66 = 114°

Since this angle is greater than 90°, the triangle is also an obtuse triangle.

Check your work.

There are 180° in a triangle, so we can check that the third angle is 114° by finding the sum of the measures of the three angles: 33 + 33 + 114 = 180°.

PRACTICE LAP

1. If angles 1 and 2 are alternating angles, and the measure of angle 2 is 110°, what is the measure of angle 1?

2. Angles *F* and *G* form a right angle. If angle *F* is 42°, what is the measure of angle *G*?

3. Angles 1 and 2 are supplementary, and angles 1 and 3 are vertical. If the measure of angle 2 is 56°, what is the measure of angle 3?

4. A triangle has one angle that is 13° and one angle that is 77°. What kind of triangle is it?

5. If an isosceles triangle has one angle that is 100°, what are the measures of the other two angles?

TRIANGLES: SUBJECT REVIEW

The formula for **area** of a triangle is $A = \frac{1}{2}bh$, where A is the area, b is the base of the triangle, and h is the height of the triangle. If the base of a triangle is 4 feet and the height is 5 feet, then the area of the triangle is equal to $\frac{1}{2}(4)(5) = \frac{1}{2}(20) = 10$ feet^2.

The **perimeter** of a triangle is equal to the sum of the lengths of its three sides. If a triangle is **equilateral**, then its perimeter can be found by the formula $P = 3s$, where s is the length of one side of the triangle. If a triangle is **isosceles**, then its perimeter is equal to twice the length of one of the sides that are equal plus the length of the remaining side. A triangle with sides that measure 4 inches, 8 inches, and 6 inches has a perimeter of $4 + 8 + 6 = 18$ inches.

If two triangles are **similar**, we can use the ratio of their sides or the ratio of their areas to find any missing measurements. If we know that the ratio of triangle A to triangle B is 2:1 and that the area of triangle A is 24 square units, then the area of triangle B, x, can be found by solving this proportion: $\frac{2}{1} = \frac{24}{x}$, $2x = 24$, $x = 12$. The area of triangle B is 12 square units.

If we are given the lengths of two sides of a right triangle, we can find the length of the missing side using the **Pythagorean theorem**: $a^2 + b^2 = c^2$, where a and b are the bases of the right triangle and c is the hypotenuse. If one base of a right triangle measures 3 feet and the other measures 4 feet, then $(3)^2 + (4)^2 = c^2$, $9 + 16 = c^2$, $25 = c^2$, and $c = 5$. The hypotenuse of the triangle is 5 feet.

FUEL FOR THOUGHT

Area is the amount of two-dimensional space that a surface covers. Area is always expressed in square units.

Perimeter is the distance around an area. Perimeter is always expressed in linear units.

Similar triangles have identical angles. All three corresponding angles are congruent, although the sides of the triangles may not be congruent.

The **Pythagorean theorem** states that the sum of the squares of the bases of a right triangle is equal to the square of the hypotenuse of the right triangle: $a^2 + b^2 = c^2$.

Most area word problems will contain the word *area*. Sometimes these problems give the base and height of a shape, and sometimes, they give the area and one of the two measurements and ask you for the other measurement. Perimeter word problems often contain the word *perimeter*, but a perimeter word problem may also ask you to find the *distance around* a figure.

Example

What is the area of a triangle whose base is 3 meters and whose height is 22 meters?

Read the entire word problem.
We are given the base and height of a triangle.
Identify the question being asked.
We are looking for the area of the triangle.
Underline the keywords and words that indicate formulas.
The phrase *area of a triangle* tells us that we must use the area formula for triangles.
Cross out extra information and translate words into numbers.
There is no extra information in this problem.
List the possible operations.
We are given the base and the height. The area of a triangle is the product of $\frac{1}{2}$, the base, and the height, so we must multiply.
Write number sentences for each operation.
$\frac{1}{2}(3)(22)$
Solve the number sentences and decide which answer is reasonable.
$\frac{1}{2}(3)(22) = \frac{1}{2}(66) = 33$ square meters
Check your work.
Since we used multiplication to find our answer, we can use division to check it. Divide the area by the base, and that value should equal half the height: $\frac{33}{3} = 11$, which is half of 22.

INSIDE TRACK

REMEMBER, YOU CAN always rewrite a formula by moving terms from one side of the equation to the other. If you are given the area and height of a triangle and you are looking for the base, you can rewrite the formula $A = \frac{1}{2}bh$ by dividing both sides of $\frac{1}{2}h$. The base, b, is equal to $\frac{2A}{h}$.

Example

If an equilateral triangle has a perimeter of 111 inches, what is the length of one side of the triangle?

Read the entire word problem.
We are given the perimeter of an equilateral triangle.
Identify the question being asked.
We are looking for the length of one side of the triangle.
Underline the keywords and words that indicate formulas.
The word *perimeter* tells us that we have the distance around a triangle, and the phrase *equilateral triangle* tells us that every side of the triangle is equal in measure. These words together mean that we can use the formula for perimeter of an equilateral triangle, $P = 3s$.
Cross out extra information and translate words into numbers.
There is no extra information in this problem.
List the possible operations.
Rewrite the formula for perimeter of an equilateral triangle to solve for s, the side of the triangle: $s = \frac{P}{3}$
Write number sentences for each operation.
Divide the perimeter by 3:
$\frac{111}{3}$
Solve the number sentences and decide which answer is reasonable.
$\frac{111}{3} = 37$ inches
Check your work.

Since we used division to find our answer, we can use multiplication to check it. The perimeter of the equilateral triangle should be equal to three times the length of one side: $(3)(37) = 111$ inches.

Almost all word problems that involve similar triangles will include the word *similar*, but remember that similar triangles are two triangles whose angles are identical. If a word problem describes two triangles as having identical angles, be prepared to set up a ratio to find a missing side or area of one of the triangles. If a word problem is about similar triangles, draw a picture to help you match up corresponding sides.

Example

Triangles *ABC* and *DEF* are similar. If the length of side *AB* is 12 feet, the length of side *DE* is 30 feet, and the length of *BC* is 8 feet, what is the length of side *EF*?

Since this is a similar triangle word problem, we'll solve it by drawing a picture. Start by drawing two triangles. It's not important how they look, but it is important how we label them. We must be sure to match angle *A* with angle *D*, angle *B* with angle *E*, and angle *C* with angle *F*:

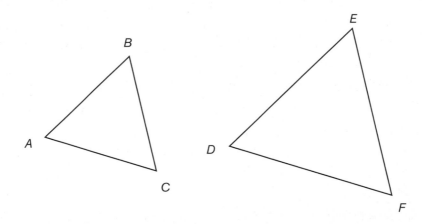

Next, label *AB* 12 feet, *DE* 30 feet, and *BC* 8 feet:

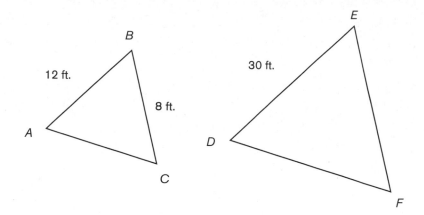

We can see now that sides *AB* and *DE* are corresponding and sides *BC* and *EF* are corresponding. Since the triangles are similar, the ratio of side *AB* to side *DE* is equal to the ratio of side *BC* to side *EF*: $\frac{12}{30} = \frac{8}{EF}$. Cross multiply and divide: $12EF = 240$, $EF = 20$ feet.

We can identify a problem that involves a right triangle if we see the word *hypotenuse*. The word problem may also state that the triangle is a right triangle, or it may simply state that the triangle contains a 90° or right angle. If a problem asks for the missing side of a right triangle, you will likely need to use the Pythagorean theorem.

Example

The legs of triangle *GHI* form a right angle. If the length of one leg is 15 feet and the length of the other leg is 20 feet, what is the length of the hypotenuse?

Read the entire word problem.
We are given the lengths of the legs of a triangle that has a right angle.
Identify the question being asked.
We are looking for the length of the hypotenuse.
Underline the keywords and words that indicate formulas.
This problem contains many keywords. The legs of the triangle form a *right angle*, which means that this is a right triangle. The word *hypotenuse* also tells us that this is a right triangle, and in order to

find the length of the hypotenuse, we will have to use the
Pythagorean theorem: $a^2 + b^2 = c^2$.

Cross out extra information and translate words into numbers.

There is no extra information in this problem.

List the possible operations.

We will need to square the legs and add those squares. Then, we will
need to take the square root of that sum.

Write number sentences for each operation.

First, square the legs and add them:

$(15)^2 + (20)^2$

Solve the number sentences and decide which answer is reasonable.

$(15)^2 + (20)^2 = 225 + 400 = 625$

Write number sentences for each operation.

The answer 625 is the square of the hypotenuse, so to find the length
of the hypotenuse, we must take the square root of 625:

$\sqrt{625}$

Solve the number sentences and decide which answer is reasonable.

$\sqrt{625} = 25$ feet

The hypotenuse of triangle *GHI* is 25 feet.

Check your work.

The square of the hypotenuse should equal the sum of the squares of
the legs: $(15)^2 + (20)^2 = (25)^2$, $225 + 400 = 625$, $625 = 625$.

INSIDE TRACK

ONLY A RIGHT triangle has a hypotenuse, so even if you are not told
that a triangle is right, if you are asked to find the length of the
hypotenuse, you are working with a right triangle, and you will need
to use the Pythagorean theorem.

6. What is the base of a triangle whose area is 48 square units and whose height is 10 units?

7. What is the perimeter of an isosceles triangle whose bases measure 12 inches and whose longest side is 20 inches?

8. The angles of triangles *JKL* and *MNO* are identical. If the area of triangle *JKL* is four times the area of triangle *MNO*, and side *JK* is 16 meters long, how long is side *MN*?

9. Right triangle *XYZ* has a hypotenuse of 10 centimeters. If the length of one leg is 6 centimeters, what is the length of the other leg?

10. If the length of one leg of an isosceles triangle *ABC* is 8 feet, what is the length of the hypotenuse of triangle *ABC*?

POLYGONS AND QUADRILATERALS: SUBJECT REVIEW

The sum of the **interior angles** of a **polygon** is equal to $180(s - 2)$, where s is the number of sides of the polygon. If a polygon has ten sides, then the sum of its interior angles is $180(10 - 2) = 180(8) = 1,440°$.

There are six major types of **quadrilaterals**: parallelograms, rhombi, rectangles, squares, trapezoids, and isosceles trapezoids. By analyzing the properties of a quadrilateral, we can determine if the quadrilateral is a **parallelogram**, **rhombus**, **rectangle**, **square**, **trapezoid**, **isosceles triangle**—or just a quadrilateral.

FUEL FOR THOUGHT

An **interior angle** is an angle found within a closed figure.

A **polygon** is a closed figure made up of line segments.

A **quadrilateral** is a polygon with four sides.

A **parallelogram** is a quadrilateral whose opposite sides are parallel and congruent. Its opposite angles are also congruent, consecutive angles are supplementary, and its diagonals bisect each other.

A **rectangle** is a parallelogram with four congruent, 90° angles and congruent diagonals.

A **rhombus** is a parallelogram with four congruent sides and perpendicular diagonals that bisect their angles.

A **square** has all of the properties of a rhombus and a rectangle: four congruent sides, four congruent, 90° angles, two pairs of parallel sides, consecutive angles that are supplementary, and diagonals that are congruent, perpendicular, bisect each other, and bisect their angles.

A **trapezoid** is a quadrilateral with at least one pair of parallel sides.

An **isosceles trapezoid** is a trapezoid with a pair of congruent sides, angles, and diagonals.

The formula for the area of any parallelogram is $A = bh$, where A is the area, b is the base of the parallelogram, and h is the height of the parallelogram. If the base of a parallelogram is 10 centimeters and the height is 40 centimeters, then the area is $(10)(40) = 400$ centimeters^2.

This formula can also be used to find the area of a rectangle, since a rectangle is a type of parallelogram, but the area of a rectangle is usually described as $A = lw$, where A is the area, l is the length of the rectangle, and w is the width of the rectangle. If the length of a rectangle is 1.2 millimeters and the width is 2.5 millimeters, then the area is $(1.2)(2.5) = 3$ millimeters^2.

The area of a square could be found with either formula, since a square is both a parallelogram and a rectangle, but it is easier to describe the area of a square as $A = s^2$, where s is the length of one side of the square, since all four sides of a square are congruent. If one side of a square is 0.17 meters, then the area of the square is $(0.17)^2 = 0.0289$ meters^2.

The area of a trapezoid is $A = \frac{1}{2}(b_1 + b_2)h$, where A is the area, b_1 and b_2 are the bases of the trapezoid, and h is the height of the trapezoid. If the lengths of the bases of a trapezoid are 3 feet and 9 feet, respectively, and the height is 10 feet, then the area of the trapezoid is $\frac{1}{2}(3 + 9)10 = \frac{1}{2}(12)(10) = (6)(10) = 60$ feet^2.

The perimeter of any quadrilateral is the sum of the lengths of each of its four sides, but we also have separate formulas for rectangles and squares. The perimeter of a rectangle can be described as $P = 2l + 2w$, since the two lengths of the rectangle are the same, as are the two widths. Since all four sides of a square measure the same, the perimeter of a square can be described as $P = 4s$. Just multiply the length of one side by 4. If one side of a square is 5.4 centimeters, then the perimeter of the square is $(5.4)(4) = 21.6$ centimeters.

Just as with triangles, quadrilateral word problems about area or perimeter usually contain the word *area* or the word *perimeter*. Phrases such as *distance around* can describe perimeter, and phrases such as *surface covered* can indicate area.

Example

What is the distance around a rectangle that has a length of 2 yards and a width of 8.9 yards?

Read the entire word problem.
We are given the length and width of a rectangle.
Identify the question being asked.
We are looking for the distance around the rectangle.
Underline the keywords and words that indicate formulas.
The words *distance around* indicate that we are looking for the perimeter of the rectangle, which is equal to $2l + 2w$.
Cross out extra information and translate words into numbers.
There is no extra information in this problem.
List the possible operations.
We will need to multiply both the length and the width by 2 and then add those products.
Write number sentences for each operation.
$2(2) + 2(8.9)$
Solve the number sentences and decide which answer is reasonable.
$2(2) + 2(8.9) = 4 + 17.8 = 21.8$ yards
Check your work.
Subtract the length of each side from the perimeter. Since the perimeter is the total distance around the rectangle, we should be left with zero after subtracting every length: $21.8 - 2 - 2 - 8.9 - 8.9 = 0$

Word problems can combine more than one formula. For instance, you may use the perimeter of a square to find the length of one side of that square, and then use the length of the side to find the area of the square.

Example

If the perimeter of a square is 44 inches, what is the area of the square?

Read the entire word problem.
We are given the perimeter of a square.
Identify the question being asked.
We are looking for the area of the square.
Underline the keywords and words that indicate formulas.
The words *perimeter* and *area* are keywords.
Cross out extra information and translate words into numbers.
There is no extra information in this problem.
List the possible operations.
The perimeter of a square is equal to 4s, so we will need to divide 44 by 4 to find the length of one side of the square. The area of a square is equal to s^2, so we will need to square our quotient.
Write number sentences for each operation.
First, find the length of one side of the square:
$\frac{44}{4}$
Solve the number sentences and decide which answer is reasonable.
$\frac{44}{4}$ = 11 inches
Write number sentences for each operation.
One side of the square is 11 inches, which means that the area of the square is equal to 11^2.
Solve the number sentences and decide which answer is reasonable.
11^2 = 121 square inches
Check your work.
Take the square root of the area to find the length of one side of the square. Then, multiply that by 4 to find the perimeter of the square, which should equal 44 inches: $\sqrt{121}$ = 11, (4)(11) = 44 inches.

CAUTION

WHEN WORKING WITH a word problem that involves both area and perimeter, be sure watch your units carefully. Perimeter is found in linear units, and area is found in square units.

PACE YOURSELF

USING A RULER, a pencil, and paper, draw a rectangle. What is the area and perimeter of your rectangle? Now, divide your rectangle into two triangles. What is the area and the perimeter of each triangle?

PRACTICE LAP

11. If a quadrilateral has four right angles, what type of quadrilateral could it be?

12. What quadrilateral has perpendicular diagonals that are not congruent?

13. One base of a trapezoid measures 16 meters, and the other measures 10 meters. If the height of the trapezoid is 4 meters, what is the area?

14. If a rhombus has an area of 117 centimeters^2 and a height of 9 centimeters, what is the base of the rhombus?

15. The area of a square is 81 square feet. If the length of each side of the square is doubled, what is the area of the new square?

CIRCLES: SUBJECT REVIEW

The formula for area of a **circle** is $A = \pi r^2$, where A is the area of the circle and r is the **radius** of the circle. If a circle has a radius of 10 feet, then its area is $\pi(10)^2 = 100\pi$ feet^2.

FUEL FOR THOUGHT

A **circle** is a set of connected points that are all the same distance from a single point, which is the center of the circle. A circle is a closed figure, but not a polygon.

A **radius** is the distance from the center of a circle to a point on the circle.

A **diameter** is the distance from one side of a circle to the other through the center of the circle.

The **circumference** of a circle is the distance around the outside of a circle.

The diameter of a circle is twice the radius, which means that $d = 2r$ and $r = \frac{1}{2}d$, where d is the diameter and r is the radius. If the diameter of a circle is 27 feet, then the radius of the circle is $\frac{1}{2}(2) = 13.5$ feet. If the radius of another circle is 2.8 centimeters, then the diameter of that circle is $2(2.8) = 5.6$ centimeters.

The circumference of a circle is equal to $2\pi r$, or πd. If the diameter of a circle is 27 feet, then the circumference of that circle is 27π feet. If the circumference of a circle is 10π inches, then the diameter of the circle is 10 inches and the radius is 5 inches.

Circumference sounds a lot like perimeter, and it may appear in word problems the same way. If a word problem asks you for the distance around a circle, it's asking you for the circumference of the circle. The keywords *radius* and *diameter* will also indicate that you are working with a circle, although the word *circle* almost always appears in any word problem that involves a circle.

SINCE WE HAVE two formulas for the circumference of a circle, one that uses the radius and one that uses the diameter, choose which formula to use based on the information given to you in a problem. If you are given the radius, use $C = 2\pi r$, and if you are given the diameter, use $C = \pi d$. You may save yourself the step of finding the radius from the diameter, or vice versa.

Example

What is the radius of a circle whose circumference is 24π units?

Read the entire word problem.
We are given a circumference.
Identify the question being asked.
We are looking for a radius.
Underline the keywords and words that indicate formulas.
The words *circumference* and *radius* are keywords.
Cross out extra information and translate words into numbers.
There is no extra information in this problem.
List the possible operations.
The formula for circumference is $C = 2\pi r$, so we can find the radius by dividing the circumference by 2π.
Write number sentences for each operation.
$\frac{24}{2\pi}$
Solve the number sentences and decide which answer is reasonable.
$\frac{24}{2\pi} = 12$ units
Check your work.
Since $C = 2\pi r$, multiply the radius by 2π to check that it is equal to the circumference, 24π units: $2\pi(12) = 24\pi$ units.

16. What is the diameter of a circle whose radius is 100 inches?

17. If the radius of a circle is 4 feet, what is the area of the circle?

18. The circumference of a circle is 18π units. What is the diameter of the circle?

19. If the area of a circle is 144π square units, what is the diameter of the circle?

20. If the area of a circle is 361π centimeters^2, what is the circumference of the circle?

VOLUME: SUBJECT REVIEW

The formula of the volume of a rectangular prism is $V = lwh$, where V is the volume of the prism, l is the length, w is the width, and h is the height. If a rectangular prism has a length of 6 inches, a width of 10 inches, and a height of 20 inches, then it has a volume of $(6)(10)(20) = 1{,}200$ inches^3.

FUEL FOR THOUGHT

VOLUME IS THE amount of space taken up by a three-dimensional object. Volume is always found in cubic units.

The formula of the volume of a cube, which is a type of rectangular prism, is $V = e^3$, where V is the volume of the prism and e is the length of one edge of the cube. A cube with an edge of 5 feet has a volume of $(5)^3 = 125$ feet^3.

The formula to find the volume of a pyramid is $V = \frac{1}{3}bh$, where V is the volume of the pyramid, b is the area of the base, and h is the height. If the pyramid has a base of 3 square feet and a height of 12 feet, then it has a volume of $\frac{1}{3}(3)(12) = 12$ feet^3.

The formula to find the volume of a cylinder is $V = \pi r^2 h$, where V is the volume of the cylinder, r is the radius, and h is the height. If a cylinder has a radius of 7 centimeters and a height of 15 centimeters, then it has a volume of $\pi(7)^2(15) = 735$ centimeters^3.

The formula for the volume of a cone is $V = \frac{1}{3}\pi r^2 h$, where V is the volume of the cone, r is the radius, and h is the height. If a cone has a radius of 15 m and a height of 30 meters, then it has a volume of $\frac{1}{3}\pi(15)^2(30) = 2{,}250$ meters^3.

The formula for the volume of a sphere is $V = \frac{4}{3}\pi r^3$, where V is the volume of the sphere and r is the radius. If a sphere has a radius of 3 inches, then it has a volume of $\frac{4}{3}\pi(3)^3 = 36\pi$ inches^3.

No tricks to a volume word problem—volume word problems almost always use the word *volume* and give the measurements of the type of solid whose volume you must find. Since volume is the amount of space an object occupies, you may be asked for the *amount of space taken up* instead of the volume of a solid. That phrase should always signal volume.

Example

If a cube has an edge of 5 centimeters, how much space does the cube take up?

Read the entire word problem.
We are given the edge of a cube.
Identify the question being asked.
We are looking for how much space it takes up.
Underline the keywords and words that indicate formulas.
The phrase *how much space does the cube take up* asks us to find the volume of a cube.
Cross out extra information and translate words into numbers.
There is no extra information in this problem.
List the possible operations.
The formula for volume of a cube is $V = e^3$, so we can find the volume by cubing, or multiplying by itself three times, the edge of the cube.
Write number sentences for each operation.
5^3
Solve the number sentences and decide which answer is reasonable.
$5^3 = 125$ centimeters^3
Check your work.

Since $V = e^3$, we can find the length of the edge by taking the cube root of the volume. The cube root of 125 should equal 5 centimeters: $3\sqrt{125}$ = 5 centimeters.

CAUTION

VOLUME WORD PROBLEMS may be easy to identify, but be sure to select the right volume formula. The volume formulas for pyramids, cylinders, cones, and spheres are very similar, but are, in fact, a little different from one another. Determine which solid you are working with, and then carefully apply the right formula.

PRACTICE LAP

21. If a rectangular prism has a length of 14 inches, a width of 9 inches, and a height of 10 inches, what is the volume of the prism?
22. A sphere has a volume of 972π centimeters³. What is the radius of the sphere?
23. What is the volume of a cylinder whose radius is 8 feet and whose height is 12 feet?
24. If a cone has a volume of 300π inches³ and a height of 36 inches, what is the radius of the cone?
25. The area of one side of a cube is 2.56 meters². What is the volume of the cube?

PACE YOURSELF

FIND A CUBE, a rectangular prism, and a cylinder in your home. A box of tissues may be a cube, a shoebox may be a rectangular prism, and a roll of paper towels may be a cylinder. Measure each with a ruler and find the volume of each. Which object has the largest volume?

SURFACE AREA: SUBJECT REVIEW

We can also find the **surface area** of solid figures.

SURFACE AREA IS the sum of the areas of each face, or side, of a solid figure.

The surface area of a rectangular prism is $SA = 2(lw + wh + lh)$, where SA is the surface area, l is the length, w is the width, and h is the height. If the length of a prism is 1.5 meters, the width is 2.8 meters, and the height is 0.5 meters, then the surface area is equal to $2[(1.5)(2.8) + (2.8)(0.5) + (1.5)(0.5)] = 2(4.2 + 1.4 + 0.75) = 2(6.35) = 12.7$ meters^2.

The surface area of a cube is $SA = 6s^2$, where SA is the surface area and s is the length of one side of the cube. If one side of the cube measures 14 centimeters, then the surface area of the cube is $6(14)^2 = 6(196) = 1,176$ centimeters^2.

The surface area of a cylinder is $SA = 2\pi r^2 + 2\pi rh$, where SA is the surface area, r is the radius, and h is the height. A cylinder with a radius of 8 millimeters and a height of 4 millimeters has a surface area of $2\pi(8)^2 + 2\pi(8)(4) = 2\pi(64) + 2\pi(32) = 128\pi + 64\pi = 192\pi$ millimeters^2.

The surface area of a sphere is $SA = 4\pi r^2$, where SA is the surface area and r is the radius. A sphere with a radius of 2 feet has a surface area of $4\pi(2)^2 = 4\pi(4) = 16\pi$ feet^2.

A word problem may describe the surface area as *the area of every face of a solid* or *the total area of a solid*. We will plug the given values into one of the four surface area formulas to find the surface area of the given solid. We may be given the area of one or more sides of the solid, or the volume of the solid. We can use those values to find the dimensions of the solid, and then find the surface area of the solid.

Example

If the area of one side of a cube is 80 centimeters², what is the total area of the cube?

Read the entire word problem.
We are given the area of one side of a cube.
Identify the question being asked.
We are looking for the total area of the cube.
Underline the keywords and words that indicate formulas.
The phrase *total area* means that we are looking for the surface area of the cube.
Cross out extra information and translate words into numbers.
There is no extra information in this problem.
List the possible operations.
We are given the area of one side of a cube, and a cube has six sides.
The surface area of the cube is equal to six times the area of one side.
Write number sentences for each operation.
(6)(80)
Solve the number sentences and decide which answer is reasonable.
(6)(80) = 480 centimeters²
Check your work.
Divide the surface area by 6, and we should get the given area of one side of the cube, 80 centimeters²: $\frac{480}{6}$ = 80 centimeters².

INSIDE TRACK

THE FORMULAS FOR volume and surface area of some solids can be complex. Remember that the exponent of the radius is always two in surface area formulas and three in volume formulas.

26. What is the surface area of a cylinder with a radius of 6 yards and a height of 18 yards?

27. The surface area of a sphere is 213.16π millimeters^3. What is the radius of the sphere?

28. If the volume of a cube is 343 inches^3, what is the surface area of the cube?

29. A rectangular prism has a length of 3 meters and a width of 2.2 meters. If the prism has a volume of 27.06 meters^3, what is the surface area of the prism?

30. The area of the base of a cylinder is 16π centimeters^2. If the height of the cylinder is 8 centimeters, what is the surface area of the cylinder?

SUMMARY

MOST GEOMETRY PROBLEMS give you a diagram, but now you're prepared to handle a geometry problem made up only of words. And at the same time, we've reviewed many important geometry rules and formulas. You've made it to the end—and you're ready to take the posttest!

ANSWERS

1. *Read the entire word problem.*

We are given the measure of one of two alternating angles.

Identify the question being asked.

We are looking for the measure of the other alternating angle.

Underline the keywords and words that indicate formulas.

The word *alternating* tells us that angles 1 and 2 are congruent—their measures are the same.

Cross out extra information and translate words into numbers.

There is no extra information in this problem.

List the possible operations.

We don't need to perform any operations or write any number sentences. Alternating angles are congruent, so since angle 2 is 110°, angle 1 is 110°.

Check your work.

We didn't perform any operations, so we don't have any work to check. Alternating angles are also known as corresponding angles, and the definition of corresponding angles is two or more congruent angles that have similar positions in a shape or figure. Our answer is correct.

2. *Read the entire word problem.*

We are given the measure of angle *F* and told that it forms a right angle with angle *G*.

Identify the question being asked.

We are looking for the measure of angle *G*.

Underline the keywords and words that indicate formulas.

The phrase *right angle* tells us that angles *F* and *G* are complementary—their measures sum to 90°.

Cross out extra information and translate words into numbers.

There is no extra information in this problem.

List the possible operations.

Since angles *F* and *G* are complementary, we must subtract the measure of angle *F* from 90° to find the measure of angle *G*.

Write number sentences for each operation.

90 – 42

Solve the number sentences and decide which answer is reasonable.

90 – 42 = 48°

Check your work.

There are 90° in a right angle, so the measures of angles *F* and *G* should add up to 90: 42 + 48 = 90°.

3. *Read the entire word problem.*

We are given the measure of angle 2 and told that angles 1 and 2 are supplementary. We are also told that angles 1 and 3 are vertical.

Identify the question being asked.

We are looking for the measure of angle 3.

Underline the keywords and words that indicate formulas.

The word *supplementary* tells us that the measures of angles 1 and 2 add up to 180°, and the word *vertical* tells us that angles 1 and 3 are congruent.

Cross out extra information and translate words into numbers.

There is no extra information in this problem.

List the possible operations.

Since angles 1 and 2 are supplementary, we must subtract the measure of angle 2 from 180° to find the measure of angle 1.

Write number sentences for each operation.

180 − 56

Solve the number sentences and decide which answer is reasonable.

180 − 56 = 124°

Angle 1 is 124°. Angles 1 and 3 are vertical angles, so angle 3 is also 124°.

Check your work.

We can check that angle 1 is 124° by adding its measure to the measure of angle 2, since these angles are supplementary: 124 + 56 = 180°. Vertical angles are defined as congruent angles that are formed by the intersection of two lines, so angle 3 does, in fact, equal angle 1.

4. *Read the entire word problem.*

We are given the measures of two angles of a triangle.

Identify the question being asked.

We are looking for the name of this kind of triangle.

Underline the keywords and words that indicate formulas.

There are no keywords in this problem.

Cross out extra information and translate words into numbers.

There is no extra information in this problem.

List the possible operations.

The two angle measures are not 60° and they are not equal, so this triangle is not equilateral or isosceles. We must add the two angles together and subtract from 180° to find the measure of the third angle. Then, we will be able to determine if the triangle is acute, obtuse, or right.

Write number sentences for each operation.

First, add the two given angles:

13 + 77

Solve the number sentences and decide which answer is reasonable.

13 + 77 = 90°

Subtract this sum from 180°, the number of degrees in a triangle, to find the measure of the third angle.

Write number sentences for each operation.

180 – 90

Solve the number sentences and decide which answer is reasonable.

180 – 90 = 90°

The third angle of the triangle is a right angle, which means that this is a right triangle.

Check your work.

We can check that the three angles add to 180°; 13 + 77 + 90 = 180°, so the third angle is, in fact, a right angle, and this triangle is a right triangle.

5. *Read the entire word problem.*

We are given the measure of one angle of an isosceles triangle.

Identify the question being asked.

We are looking for the measures of the other two angles.

Underline the keywords and words that indicate formulas.

The keyword *isosceles* tells us that two angles in the triangle are congruent.

Cross out extra information and translate words into numbers.

There is no extra information in this problem.

List the possible operations.

The given angle is 100°. It cannot be one of the base angles of the triangle, because there cannot be two obtuse angles in a triangle (since there are only 180° in a triangle). The other two angles must be the base angles. Subtract the measure of the given angle from 180° to find the total measure of the other two angles. Then, divide that difference by 2 to find the measure of each of those angles.

Write number sentences for each operation.

First, subtract 100 from 180:

180 – 100

Solve the number sentences and decide which answer is reasonable.

180 – 100 = 80°

The two base angles combined equal 80°. Divide 80 by 2 to find the measure of each base angle.

Write number sentences for each operation.

$\dfrac{80}{2}$

Solve the number sentences and decide which answer is reasonable.

$\frac{80}{2} = 40°$

The other two angles each measure 40°.

Check your work.

We can check that the three angles add up to 180°: 100 + 40 + 40 = 180°. The three angles total 180°, and two of the angles are the same in measure, so our answer is correct.

6. *Read the entire word problem.*

We are given the area and the height of a triangle.

Identify the question being asked.

We are looking for its base.

Underline the keywords and words that indicate formulas.

The keywords *base*, *area*, *height*, and *triangle* tell us that we need to use the formula for area of a triangle.

Cross out extra information and translate words into numbers.

There is no extra information in this problem.

List the possible operations.

We are given the area and the height. We can rewrite the formula $A = \frac{1}{2}bh$ by dividing both sides of $\frac{1}{2}h$. The base, b, is equal to $\frac{2A}{h}$.

Write number sentences for each operation.

First, multiply the area by 2:

2(48)

Solve the number sentences and decide which answer is reasonable.

2(48) = 96 square units

Write number sentences for each operation.

Now, divide that by the height:

$\frac{96}{10}$

Solve the number sentences and decide which answer is reasonable.

$\frac{96}{10} = 9.6$ units

The base of the triangle is 9.6 units.

Check your work.

Since $A = \frac{1}{2}bh$, check that half the product of the base and the height equals the area, 48 square units: $\frac{1}{2}(9.6)(10) = \frac{1}{2}(96) = 48$ square units.

7. *Read the entire word problem.*

We are given the measures of the bases and the longest side of an isosceles triangle.

Identify the question being asked.

We are looking for the perimeter of the triangle.

Underline the keywords and words that indicate formulas.

We are asked to find the perimeter, and the keywords *isosceles triangle* tell us that the base angles of the triangle are equal. We have been given all three sides of the triangle, and we must add them to find the perimeter.

Cross out extra information and translate words into numbers.

There is no extra information in this problem.

List the possible operations.

The perimeter of a triangle is equal to the sum of the lengths of the sides, so we must add.

Write number sentences for each operation.

14 + 14 + 20

Solve the number sentences and decide which answer is reasonable.

14 + 14 + 20 = 44 inches

The perimeter of the triangle is 44 inches.

Check your work.

Since the triangle is isosceles, if we subtract base lengths from the perimeter, we should be left with the other side of the triangle, which is 20 inches long: 44 − 12 − 12 = 20 inches.

8. We are told that the angles of triangles *JKL* and *MNO* are identical, which means that they are similar triangles. Draw and label triangles *JKL* and *MNO*:

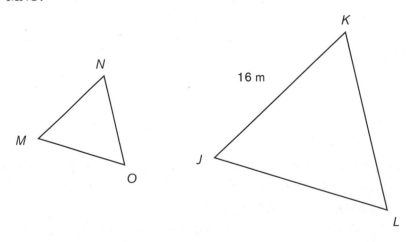

We can use a proportion to find the length of side *MN*. Since the area of triangle *JKL* is four times the area of triangle *MNO*, the ratio of their sides is 4:1; $\frac{4}{1} = \frac{16}{MN}$. Cross multiply and divide: $4MN = 16$, $MN = 4$ meters.

9. *Read the entire word problem.*

We are given the measures of one leg and the hypotenuse of a right triangle.

Identify the question being asked.

We are looking for the measure of the other leg.

Underline the keywords and words that indicate formulas.

We are told we are working with a right triangle, which the word *hypotenuse* also indicates.

Cross out extra information and translate words into numbers.

There is no extra information in this problem.

List the possible operations.

Since we have a right triangle and the lengths of two sides, we can use the Pythagorean theorem to find the length of the missing side. Rewrite the formula $a^2 + b^2 = c^2$ as $b^2 = c^2 - a^2$, since we are looking for the length of a leg.

Write number sentences for each operation.

First, find the difference between the square of the hypotenuse and the square of one leg:

$(10)^2 - (6)^2$

Solve the number sentences and decide which answer is reasonable.

$(10)^2 - (6)^2 = 100 - 36 = 64$

Write number sentences for each operation.

The square of the leg is 64, so to find the length of the leg, we must take the square root of 64: $\sqrt{64}$

Solve the number sentences and decide which answer is reasonable.

$\sqrt{64} = 8$

The length of the other leg is 8 centimeters.

Check your work.

The square of the hypotenuse should equal the sum of the squares of the legs: $(6)^2 + (8)^2 = (100)^2$, $36 + 64 = 100$, $100 = 100$.

10. *Read the entire word problem.*

We are given the measures of one leg of a triangle.

Identify the question being asked.

We are looking for the measure of the hypotenuse.

Underline the keywords and words that indicate formulas.

The word *hypotenuse* indicates we are working with a right triangle. Since the triangle is also isosceles, it must be an isosceles right triangle.

Cross out extra information and translate words into numbers.

There is no extra information in this problem.

List the possible operations.

Since the triangle is isosceles, both legs are 8 feet. Use the formula $a^2 + b^2 = c^2$ to find c, the length of the hypotenuse. We will need to square the legs and add those squares. Then, we will need to take the square root of that sum.

Write number sentences for each operation.

First, square the legs and add them:

$(8)^2 + (8)^2$

Solve the number sentences and decide which answer is reasonable.

$(8)^2 + (8)^2 = 64 + 64 = 128$

Write number sentences for each operation.

The square of the hypotenuse is 128, so to find the length of the hypotenuse, we must take the square root of 128:

$\sqrt{128}$

Solve the number sentences and decide which answer is reasonable.

$\sqrt{128} = 8\sqrt{2}$ feet.

The hypotenuse of triangle ABC is $8\sqrt{2}$ feet.

Check your work.

The square of the hypotenuse should equal the sum of the squares of the legs: $(8)^2 + (8)^2 = (8\sqrt{2})^2$, $64 + 64 = 128$.

11. *Read the entire word problem.*

We are told that a quadrilateral has four right angles.

Identify the question being asked.

We are looking for the type of quadrilateral that it could be.

Underline the keywords and words that indicate formulas.

The keywords *four right angles* tell us that this quadrilateral is either a rectangle or a square.

Cross out extra information and translate words into numbers.

There is no extra information in this problem.

List the possible operations.

There are no operations to perform. A rectangle has four right angles, and a square is a type of rectangle, so this quadrilateral could be a square or a rectangle.

Check your work.

The definition of a rectangle is a parallelogram with four congruent, 90° angles and congruent diagonals, and a square is a rectangle, so this answer is correct.

12. *Read the entire word problem.*

We are told that a quadrilateral has perpendicular diagonals that are not congruent.

Identify the question being asked.

We are looking for the type of quadrilateral that it could be.

Underline the keywords and words that indicate formulas.

The keywords *perpendicular diagonals* tell us that this quadrilateral is a rhombus or a square, and the keywords *not congruent* tell us that it is not a square, since the diagonals of a square are congruent.

Cross out extra information and translate words into numbers.

There is no extra information in this problem.

List the possible operations.

There are no operations to perform. A rhombus has perpendicular diagonals, as does a square since a square is a type of rhombus. But a square has congruent diagonals, so this quadrilateral must be a rhombus.

Check your work.

The definition of a rhombus is a parallelogram with four congruent sides and perpendicular diagonals that bisect their angles, while a square has all of the properties of a rhombus and a rectangle, and the diagonals of a rectangle are congruent.

13. *Read the entire word problem.*

We are given the measures of the bases and height of a trapezoid.

Identify the question being asked.

We are looking for the area of the trapezoid.

Underline the keywords and words that indicate formulas.

The words *area* and *trapezoid* indicate that we must use the formula for area of a trapezoid: $A = \frac{1}{2}(b_1 + b_2)h$.

Cross out extra information and translate words into numbers.

There is no extra information in this problem.

List the possible operations.

First, we must add the bases and multiply by one-half. Then, we must multiply that product by the height.

Write number sentences for each operation.

$\frac{1}{2}(16 + 10)$

Solve the number sentences and decide which answer is reasonable.

$\frac{1}{2}(16 + 10) = \frac{1}{2}(26) = 13$ meters

Write number sentences for each operation.

Now multiply that product by the height, 4:

$(13)(4)$

Solve the number sentences and decide which answer is reasonable.

$(13)(4) = 52$ meters

The area of the trapezoid is 52 meters.

Check your work.

Divide the area by half the sum of the bases. This quotient should equal the height, 4 meters: $\dfrac{52}{[(\frac{1}{2})(16 + 10)]} = \dfrac{52}{[(\frac{1}{2})(26)]} = \dfrac{52}{13} = 4$ meters.

14. *Read the entire word problem.*

We are given the area and height of a rhombus.

Identify the question being asked.

We are looking for its base.

Underline the keywords and words that indicate formulas.

The words *area* and *rhombus* indicate that we must use the formula for area of a parallelogram, since a rhombus is a parallelogram: $A = bh$.

Cross out extra information and translate words into numbers.

There is no extra information in this problem.

List the possible operations.

Rewrite the formula to solve for b, the base of the rhombus: $b = \frac{A}{h}$.

Divide the area by the height.

Write number sentences for each operation.

$\frac{117}{9}$

Solve the number sentences and decide which answer is reasonable.

$\frac{117}{9} = 13$ centimeters

The base of the rhombus is 13 centimeters.

Check your work.

Since the area of a rhombus is equal to the product of its base and its height, multiply 13 by 9 to see if it equals the area of the rhombus, 117 centimeters^2: $(13)(9) = 117$ centimeters^2.

15. *Read the entire word problem.*

We are given the area of a square and how it changes.

Identify the question being asked.

We are looking for the new area of the square.

Underline the keywords and words that indicate formulas.

The words *area* and *square* indicate that we must use the formula for area of a square, to find the length of one side of the square: $A = s^2$.

Cross out extra information and translate words into numbers.

There is no extra information in this problem.

List the possible operations.

We can take the square root of the area to find the length of one side of the square. Then, we can double that length and use the area formula again to find the area of the new square.

Write number sentences for each operation.

First, take the square root of the area:

$\sqrt{81}$

Solve the number sentences and decide which answer is reasonable.

$\sqrt{81} = 9$ feet

Write number sentences for each operation.

One side of the original square is 9 feet. To find the length of one side of the new square, multiply 9 by 2:

$(9)(2)$

Solve the number sentences and decide which answer is reasonable.

$(9)(2) = 18$ feet

Write number sentences for each operation.

One side of the new square is 18 feet. To find the area of the new square, square 18:

18^2

Solve the number sentences and decide which answer is reasonable.

$18^2 = 324$ square feet

Check your work.

If the length of a side of a square doubles, the area of the square increases by four times. Multiply the original area, 81 square feet, by 4 to see if it equals the new area: $(81)(4) = 324$ square feet.

16. *Read the entire word problem.*

    We are given a radius.

    Identify the question being asked.

    We are looking for a diameter.

    Underline the keywords and words that indicate formulas.

    The words *diameter* and *radius* are keywords.

    Cross out extra information and translate words into numbers.

    There is no extra information in this problem.

    List the possible operations.

    The formula for diameter is $d = 2r$. Multiply the radius by 2 to find the diameter.

    Write number sentences for each operation.

    2(100)

    Solve the number sentences and decide which answer is reasonable.

    2(100) = 200 inches

    Check your work.

    Since diameter is twice the radius, divide the diameter by 2 to check that the radius is, in fact, 100 inches: $\frac{200}{100} = 2$ inches.

17. *Read the entire word problem.*

    We are given a radius.

    Identify the question being asked.

    We are looking for an area.

    Underline the keywords and words that indicate formulas.

    The words *radius* and *area* are keywords.

    Cross out extra information and translate words into numbers.

    There is no extra information in this problem.

    List the possible operations.

    The formula for the area of a circle is $A = \pi r^2$. Square the radius and multiply by π to find the area.

    Write number sentences for each operation.

    $\pi(4)^2$

    Solve the number sentences and decide which answer is reasonable.

    $\pi(4)^2 = 16\pi$ inches^2

    Check your work.

    Divide the area by π, and then take the square root of the quotient. This should give us back the radius, 4 feet: $\frac{16\pi}{\pi} = 16$, $\sqrt{16} = 4$ feet.

18. *Read the entire word problem.*

We are given a circumference.

Identify the question being asked.

We are looking for a diameter.

Underline the keywords and words that indicate formulas.

The words *circumference* and *diameter* are keywords.

Cross out extra information and translate words into numbers.

There is no extra information in this problem.

List the possible operations.

The formula for circumference is $C = \pi d$, so we can find the diameter by dividing the circumference by π.

Write number sentences for each operation.

$$\frac{18\pi}{\pi}$$

Solve the number sentences and decide which answer is reasonable.

$\frac{18\pi}{\pi} = 18$ units

Check your work.

Multiply the diameter by π. This should give us back the circumference: $(18)(\pi) = 18\pi$, so our answer is correct.

19. *Read the entire word problem.*

We are given an area.

Identify the question being asked.

We are looking for a diameter.

Underline the keywords and words that indicate formulas.

The words *area* and *diameter* are keywords.

Cross out extra information and translate words into numbers.

There is no extra information in this problem.

List the possible operations.

The formula for the area of a circle is $A = \pi r^2$, so we can find the radius by dividing the area by π, and then taking the square root of the quotient. Once we have the radius, we can multiply it by 2 to find the diameter.

Write number sentences for each operation.

Divide the area by π:

$$\frac{144\pi}{\pi}$$

Solve the number sentences and decide which answer is reasonable.

$\frac{144\pi}{\pi} = 144$ square units

Write number sentences for each operation.

Find the radius by taking the square root of 144:

$\sqrt{144}$

Solve the number sentences and decide which answer is reasonable.

$\sqrt{144} = 12$

The radius of the circle is 12 units. To find the diameter, multiply the radius by 2.

Write number sentences for each operation.

(12)(2)

Solve the number sentences and decide which answer is reasonable.

(12)(2) = 24 units

Check your work.

Divide the diameter by 2 to find the radius. Then, square the radius and multiply by π. This should give us back the area, 144π square units. $\frac{24}{2}$ = 12, $\pi(12)^2 = 144\pi$ square units.

20. *Read the entire word problem.*

We are given an area.

Identify the question being asked.

We are looking for a circumference.

Underline the keywords and words that indicate formulas.

The words *area* and *circumference* are keywords.

Cross out extra information and translate words into numbers.

There is no extra information in this problem.

List the possible operations.

The formula for the area of a circle is $A = \pi r^2$, so we can find the radius by dividing the area by π, and then taking the square root of the quotient. Once we have the radius, we can multiply it by 2π to find the circumference.

Write number sentences for each operation.

Divide the area by π:

$\frac{361\pi}{\pi}$

Solve the number sentences and decide which answer is reasonable.

$\frac{361\pi}{\pi} = 361$ square units

Write number sentences for each operation.

Find the radius by taking the square root of 361:

$\sqrt{361}$

Solve the number sentences and decide which answer is reasonable.

$\sqrt{361} = 19$

The radius of the circle is 19 units. To find the circumference, multiply the radius by 2π.

Write number sentences for each operation.

$(19)(2\pi)$

Solve the number sentences and decide which answer is reasonable.

$(19)(2\pi) = 38\pi$ units

Check your work.

Divide the circumference by 2π to find the radius. Then, square the radius and multiply by π. This should give us back the area, 361π square units. $\frac{38\pi}{2\pi} = 19$, $\pi(19)^2 = 361\pi$ square units.

21. *Read the entire word problem.*

We are given the length, width, and height of a rectangular prism.

Identify the question being asked.

We are looking for its volume.

Underline the keywords and words that indicate formulas.

The words *rectangular prism* and *volume* are keywords. We will use the formula for the volume of a rectangular prism.

Cross out extra information and translate words into numbers.

There is no extra information in this problem.

List the possible operations.

The formula for the volume of a rectangular prism is $V = lwh$, so we can multiply the three given values to find the volume.

Write number sentences for each operation.

$(14)(9)(10)$

Solve the number sentences and decide which answer is reasonable.

$(14)(9)(10) = 1,260$ inches^3

Check your work.

Divide the volume by the length and the width. This should give us back the height: $\frac{1,260}{(14)(9)} = \frac{1,260}{126} = 10$ inches.

22. *Read the entire word problem.*

We are given the volume of a sphere.

Identify the question being asked.

We are looking for its radius.

Underline the keywords and words that indicate formulas.

The words *sphere* and *volume* are keywords. We will use the formula for the volume of a sphere prism to find the radius of the sphere.

Cross out extra information and translate words into numbers.

There is no extra information in this problem.

List the possible operations.

The formula for the volume of a sphere is $V = \frac{4}{3}\pi r^3$. Divide both sides of the equation by $\frac{4}{3}\pi$ and take the cube root of both sides: $r = \sqrt[3]{\frac{3V}{4\pi}}$. First, multiply the volume by 3. Then, divide that product by 4π. Finally, take the cube root of the result.

Write number sentences for each operation.

$(3)(972\pi)$

Solve the number sentences and decide which answer is reasonable.

$(3)(972\pi) = 2{,}916\pi$ centimeters^3

Write number sentences for each operation.

Divide 2,916 by 4π:

$\frac{2{,}916\pi}{4\pi}$

Solve the number sentences and decide which answer is reasonable.

$\frac{2{,}916\pi}{4\pi} = 729$

Write number sentences for each operation.

Take the cube root of 729:

$\sqrt[3]{729}$

Solve the number sentences and decide which answer is reasonable.

$\sqrt[3]{729} = 9$ centimeters

The radius of the sphere is 9 centimeters.

Check your work.

Plug the radius into the formula for the volume of a sphere. $\frac{4}{3}\pi r^3 = \frac{4}{3}\pi(9)^3$ $= \frac{4}{3}(729)\pi = 972\pi$ centimeters^3.

23. *Read the entire word problem.*

We are given the radius and height of a cylinder.

Identify the question being asked.

We are looking for its volume.

Underline the keywords and words that indicate formulas.

The words *cylinder* and *volume* are keywords. We will use the formula for the volume of a cylinder.

Cross out extra information and translate words into numbers.

There is no extra information in this problem.

List the possible operations.

The formula for the volume of a cylinder is $V = \pi r^2 h$, so we can square the radius and multiply by the height to find the volume.

Write number sentences for each operation.

$\pi(8)^2(12)$

Solve the number sentences and decide which answer is reasonable.

$\pi(8)^2(12) = 768\pi$ feet^3

Check your work.

Divide the volume by the height and π. Then, take the square root, and we should have the radius: $\frac{768\pi}{12\pi} = 64$, $\sqrt{64} = 8$ feet.

24. *Read the entire word problem.*

We are given the volume and height of a cone.

Identify the question being asked.

We are looking for its radius.

Underline the keywords and words that indicate formulas.

The words *cone* and *volume* are keywords. We will use the formula for the volume of a cone.

Cross out extra information and translate words into numbers.

There is no extra information in this problem.

List the possible operations.

The formula for the volume of a cone is $V = \frac{1}{3}\pi r^2 h$, so we can rewrite this equation to solve for r, the radius. Multiply both sides of the equation by 3 and divide by π and h. Then, take the square root of both sides.

Write number sentences for each operation.

First, multiply the volume by 3:

$(300\pi)(3)$

Solve the number sentences and decide which answer is reasonable.

$(300\pi)(3) = 900\pi$ inches^3

Write number sentences for each operation.

Next, divide that product by π and h:

$\frac{(900\pi)}{(36\pi)}$

Solve the number sentences and decide which answer is reasonable.

$\frac{(900\pi)}{(36\pi)} = 25$ inches^2

Write number sentences for each operation.

Finally, take the square root:

$\sqrt{25}$

Solve the number sentences and decide which answer is reasonable.

$\sqrt{25}$ = 5 inches

Check your work.

Plug the radius and height into the formula for volume of a cone. $\frac{1}{3}\pi r^2 h$
= $\frac{1}{3}\pi(5)^2(36)$ = $\frac{1}{3}(25)\pi(36)$ = $(25\pi)(12)$ = 300π inches^3.

25. *Read the entire word problem.*

We are given the area of one side of a cube.

Identify the question being asked.

We are looking for the volume of the cube.

Underline the keywords and words that indicate formulas.

The words *area*, *cube*, and *volume* are keywords. We will use the formula
for the area of a square and the formula for the volume of a cone.

Cross out extra information and translate words into numbers.

There is no extra information in this problem.

List the possible operations.

We must use the area of one side of the cube to find the length of one
edge of the cube. Then, we can use that length to find the volume of the
cube. Since the formula for the area of a square is $A = s^2$, take the square
root of the area to find the length of one side, or edge. The formula for
the volume of a cube is $V = e^3$, so we must then take the length of one
edge and cube it to find the volume.

Write number sentences for each operation.

First, take the square root of the area:

$\sqrt{2.56}$

Solve the number sentences and decide which answer is reasonable.

$\sqrt{2.56}$ = 1.6 meters

Write number sentences for each operation.

Now, cube the length of one edge:

$(1.6)^3$

Solve the number sentences and decide which answer is reasonable.

$(1.6)^3$ = 4.096 meters^3

Check your work.

Take the cube root of the volume to find the length of one edge, and then
square that to find the area of one side of the cube: $3\sqrt{4.096}$ = 1.6, $(1.6)^2$
= 2.56 meters^2.

26. *Read the entire word problem.*

We are given the radius and height of a cylinder.

Identify the question being asked.

We are looking for its surface area.

Underline the keywords and words that indicate formulas.

The words *cylinder* and *surface area* are keywords. We will use the formula for the surface area of a cylinder.

Cross out extra information and translate words into numbers.

There is no extra information in this problem.

List the possible operations.

The formula for the surface area of a cylinder is $SA = 2\pi r^2 + 2\pi rh$, so we must begin by squaring the radius and multiplying it by 2π. Then, we must multiply the radius and the height by 2π. Finally, we must add those two products together.

Write number sentences for each operation.

First, square the radius and multiply it by 2π:

$[(6)^2](2\pi)$

Solve the number sentences and decide which answer is reasonable.

$[(6)^2](2\pi) = (36)(2\pi) = 72\pi$

Write number sentences for each operation.

Now, multiply the radius and the height by 2π:

$(6)(18)(2\pi)$

Solve the number sentences and decide which answer is reasonable.

$(6)(18)(2\pi) = 216\pi$

Write number sentences for each operation.

Add the two products:

$72\pi + 216\pi$

Solve the number sentences and decide which answer is reasonable.

$72\pi + 216\pi = 288\pi$

Check your work.

Plug both the radius and the height into the surface area formula and do the calculations again to insure your answer is correct: $2\pi(6)^2 + 2\pi(6)(18) = 72\pi + 216\pi = 288\pi$.

27. *Read the entire word problem.*

We are given the surface area of a sphere.

Identify the question being asked.

We are looking for its radius.

Underline the keywords and words that indicate formulas.

The words *sphere* and *surface area* are keywords. We will use the formula for the surface area of a sphere to find the radius of the sphere.

Cross out extra information and translate words into numbers.

There is no extra information in this problem.

List the possible operations.

The formula for the surface area of a sphere is $SA = 4\pi r^2$, so we can rewrite it to solve for the radius by dividing both sides by 4π, and then taking the square root of both sides.

Write number sentences for each operation.

First, divide the surface area by 4π:

$$\frac{(213.16\pi)}{(4\pi)}$$

Solve the number sentences and decide which answer is reasonable.

$$\frac{(213.16\pi)}{(4\pi)} = 53.29$$

Write number sentences for each operation.

Now, take the square root of 53.29:

$$\sqrt{53.29}$$

Solve the number sentences and decide which answer is reasonable.

$$\sqrt{53.29} = 7.3$$

Check your work.

Plug the radius into the surface area formula to find the surface area of the sphere: $4\pi r^2 = 4\pi(7.3)^2 = 4\pi(53.29) = 213.16\pi$ millimeters^3.

28. *Read the entire word problem.*

We are given the volume of a cube.

Identify the question being asked.

We are looking for its surface area.

Underline the keywords and words that indicate formulas.

The words *cube*, *volume*, and *surface area* are keywords. We will use the formula for the volume of a cube to find the edge of the cube, and then use the formula for surface area of a cube.

Cross out extra information and translate words into numbers.

There is no extra information in this problem.

List the possible operations.

The formula for the volume of a cube is $V = e^3$, so we can find the edge of the cube by taking the cube root of the volume. Since the formula for surface area of a cube is $SA = 6s^2$, once we have the length of one edge, or side, we can square it and multiply by 6 to find the surface area.

Write number sentences for each operation.

First, take the cube root of 343:

$\sqrt[3]{343}$

Solve the number sentences and decide which answer is reasonable.

$\sqrt[3]{343} = 7$ inches

Write number sentences for each operation.

Now, square that length and multiply by 6:

$6(7)^2$

Solve the number sentences and decide which answer is reasonable.

$6(7)^2 = 6(49) = 294$ inches^2

Check your work.

Divide the surface area by 6 and take the square root. This will give us the length of one edge of the cube. Cube that length to get back the volume of the cube: $\frac{294}{6} = 49$, $49^3 = 343$ inches^3.

29. *Read the entire word problem.*

We are given the length, width, and volume of a rectangular prism.

Identify the question being asked.

We are looking for its surface area.

Underline the keywords and words that indicate formulas.

The words *rectangular prism*, *volume*, and *surface area* are keywords. We will use the formula for the volume of a rectangular prism to find the height of the prism, and then use the formula for the surface area of a rectangular prism.

Cross out extra information and translate words into numbers.

There is no extra information in this problem.

List the possible operations.

The formula for the volume of a rectangular prism is $V = lwh$. We can rewrite that as $h = \frac{V}{lw}$ by dividing both sides of the equation by lw. Once we have the height of the rectangular prism, we can plug it, the length, and the width into the formula for the surface area of a rectangular prism: $SA = 2(lw + wh + lh)$.

Write number sentences for each operation.

First, divide the volume by the product of the length and width to find the height:

$$\frac{27.06}{(3)(2.2)}$$

Solve the number sentences and decide which answer is reasonable.

$$\frac{27.06}{(3)(2.2)} = \frac{27.06}{6.6} = 4.1 \text{ meters}$$

Write number sentences for each operation.

The height of the prism is 4.1 meters. Now, plug the values of the length, width, and height into the formula for the surface area of a rectangular prism:

$$2[(3)(2.2) + (2.2)(4.1) + (3)(4.1)]$$

Solve the number sentences and decide which answer is reasonable.

$$2[(3)(2.2) + (2.2)(4.1) + (3)(4.1)] = 2(6.6 + 9.02 + 12.3) = 2(27.92)$$
$$= 55.84 \text{ meters}^2$$

Check your work.

Check that the height you found, 4.1 meters, is correct by multiplying it by the length and width to see if it equals the given volume, 27.06 meters^3: $(3)(2.2)(4.1) = 27.06$ meters^3.

30. *Read the entire word problem.*

We are given the height and the area of the base of a cylinder.

Identify the question being asked.

We are looking for its surface area.

Underline the keywords and words that indicate formulas.

The words *cylinder*, *area*, and *surface area* are keywords. We will use the formula for the surface area of a cylinder.

Cross out extra information and translate words into numbers.

There is no extra information in this problem.

List the possible operations.

The formula for the surface area of a cylinder is $SA = 2\pi r^2 + 2\pi rh$. The bases of a cylinder are circles, so the first part of the surface area formula, $2\pi r^2$, is equal to twice the area of the base. We are given the area of the base, so we can multiply that by 2 to find the first part of the surface area of the cylinder. To find the radius of the cylinder, we must divide the area of one base by π and then take the square root. Once we have the radius, we can multiply it by 2π and by the height, and this will give us the second part of the surface area. To find the total surface area, we will add the two parts together.

Write number sentences for each operation.

First, multiply the area of one base by 2:

$(16\pi)(2)$

Solve the number sentences and decide which answer is reasonable.

$(16\pi)(2) = 32\pi$

Write number sentences for each operation.

We have the first part of the surface area. Next, find the radius of the cylinder. Divide the area of one base by π and then take the square root:

$$\frac{(16\pi)}{\pi}$$

Solve the number sentences and decide which answer is reasonable.

$$\frac{(16\pi)}{\pi} = 16$$

Write number sentences for each operation.

$$\sqrt{16}$$

Solve the number sentences and decide which answer is reasonable.

$\sqrt{16} = 4$ centimeters

Write number sentences for each operation.

The radius is 4 centimeters. Multiply it by 2π and by the height to find the second part of the surface area.

$(2\pi)(4)(8)$

Solve the number sentences and decide which answer is reasonable.

$(2\pi)(4)(8) = 64\pi$

Write number sentences for each operation.

Finally, add the first part of the surface area, the area of the bases, to the second part of the surface area, the area of the curved surface:

$32\pi + 64\pi$

Solve the number sentences and decide which answer is reasonable.

$32\pi + 64\pi = 96\pi$ centimeters^2

Check your work.

Plug both the radius and the height into the surface area formula:

$2\pi(4)^2 + 2\pi(4)(8) = 32\pi + 64\pi = 96\pi$ centimeters^2.

POSTTEST

Now that you have read through all the chapters in this book, you are ready to take the posttest. Like the pretest, the posttest has 50 questions, which cover the same topics presented in this book. The questions are similar to the pretest, so that you can see your progress and improvement since taking the pretest.

After completing the posttest, check your answers. Explanations are provided for every answer, along with the chapter in which the skill being tested is taught. Compare your score on the posttest to your score on the pretest. Did you improve? Did you learn from any mistakes you made in the pretest? The posttest can help you identify which areas you have mastered and which areas you need to practice. Return to the chapters that cover the topics that are tough for you until those topics become your strengths. You'll be able to ace the pretest, the posttest, and every kind of word problem.

Answer the following questions. Reduce all fractions to their simplest form.

1. Find the difference between 56 and 37.

2. Each press at Post Plant can print 250 newspapers in an hour. How many newspapers can six presses print in an hour?

3. There are 360 rooms in a hotel. If there are 15 floors in the hotel, how many rooms are on each floor?

4. If Terra has 18 dresses and Souyma has 15 dresses, how many dresses do they have altogether?

5. A hummingbird beats its wings 53 times per second. How many times does it beat its wings in a minute?

6. A science experiment requires 5 milliliters of iodine and 120 milliliters of water. If the experiment will be performed 15 times, how much iodine is needed?

7. Skate rental at Pro Rink costs $5.50, and the rink has 85 pairs of skates. If the rink orders 35 new pairs of skates, how many pairs will it have in total?

8. There are 38 people in line waiting for the doors of a theater to open. After 17 people are let in, how many are still waiting outside?

9. Dharmesh's party has 18 guests. If six more people arrive, how many people will be at the party?

10. Svetlana, Veronica, Rachel, Susie, and Cara are each planning their birthday parties. Svetlana's birthday comes before Veronica's, but after Cara's birthday. Cara's birthday is before Susie's, but after Rachel's. If each girl has her party on her birthday, which party will happen first?

11. A restaurant has booths for groups of four and tables for groups of five. If 53 people are seated at a total of 12 booths and tables, how many booths are filled and how many tables are filled?

12. Megapark has a water park, an amusement park, and a kiddie park. On Sunday, 204 people visited the water park, 370 people visited the amusement park, and 295 people visited the kiddie park. There were 620 total visitors to Megapark that day, 77 of which went only to the water park, 187 of which went only to the amusement park, and 151 of which went only to the kiddie park. How many people visited all three parks?

13. What is ten fewer than twelve?

14. Alimi catches 23 fish. If Alimi catches 12 more fish than Larry, how many fish does Larry catch?

15. Hector attends a book convention and adds 23 books to his collection. If he has 289 books now, how many did he have before the convention?

16. A fountain contains 640 ounces of water. How many ounces are left in the fountain after 45 eight-ounce cups of water are filled?

17. Ken is loading equipment onto a bus. Each knapsack weighs 12 pounds, and each duffel bag weighs 20 pounds. If Ken loads 12 knapsacks and five duffel bags onto the bus, how much weight in baggage is on the bus?

18. A tour bus holds 22 people. The bus makes eight trips per day, and the tour bus company has nine buses. How many people can the tour bus company serve in five days?

19. What is the sum of six fewer than eleven and nineteen?

20. What is the product of eight and fourteen, minus the sum of twelve and four more than nine?

21. Find an algebraic expression that is equal to seven more than the product of a number and thirteen.

22. Write an algebraic expression that is equivalent to half the square of a number minus fifteen times the number.

23. The sum of three times a number and eleven is equal to five. What is the number?

24. One fewer than nine times a number is five more than eight times the number. Find the number.

25. Negative six times a number, plus four, is less than twelve fewer than two multiplied by the number. Find the set of values that describes the number.

26. Marco's gas tank is $\frac{5}{8}$ full. Over the weekend, he uses $\frac{1}{4}$ of the gas in the tank. What fraction of the tank is full now?

27. Gino buys 2 pounds of chocolate. $\frac{3}{10}$ of the chocolate is white chocolate, and the rest is dark chocolate. If Gino shares the dark chocolate with five friends, how many pounds of chocolate does each friend receive?

28. A leaky faucet loses 0.06 ounces of water per hour. At the end of a day, how many ounces of water has it lost?

29. Christian's hard drive holds 50 gigabytes of data. He had 26.754 gigabytes of data saved, and then he deleted 8.24 gigabytes of that data. How many gigabytes of the hard drive are free?

30. Dr. Wilcox can schedule 30 appointments per week. If he has 21 appointments for this week, what percent of his schedule is booked?

31. An autographed celebrity photo normally sells for $120, but the price was increased by 15% when the celebrity's new movie was released. When the movie failed to do well, the price of the photo was reduced by 25%. What is the price of the photo now?

32. Nora's thermometer says that the temperature is 77° Fahrenheit. What is the temperature in degrees Celsius?

33. Janine puts $8,200 in a bank for three months. If she gains interest at a rate of 4.2% per year, how much interest does she earn?

34. A roller coaster travels at a rate of 65 miles per hour for four seconds. To the nearest thousandth of a mile, how far did the roller coaster travel?

35. Ida has a 20-ounce salt solution that is 5% salt. How much salt must she add to the solution to make it 50% salt?

36. How many liters of water must be evaporated from 45 liters of a 15% alcohol solution to make the concentration 75% alcohol?

37. Karen's jewelry box contains 12 necklaces and 30 pairs of earrings. What is the ratio of earrings to necklaces?

38. A hockey team has 14 left-handed players and six right-handed players. What is the ratio of left-handed players to the total number of players?

39. The ratio of cucumber slices to tomato slices in a salad is 4:3. If there are 12 cucumber slices in the salad, how many tomato slices are in the salad?

40. The ratio of first-class seats to business-class seats on a plane is 2:25. If there are 324 total seats on the plane, how many seats are in first class?

41. A mosaic contains red tiles and blue tiles. The ratio of red tiles to blue tiles is 3:8, and there are 1,692 red tiles in the mosaic. How many total tiles make up the mosaic?

42. A library records the number of visitors it has each day for two weeks: 32, 45, 67, 77, 45, 56, 21, 34, 57, 71, 57, 42, 57, and 30, respectively. What is the range of visitors the library gets over those two weeks?

43. There are six seventh-grade math classes at Sheepsend Middle School. The first class has 24 students, the second class has 27 students, the third class has 23 students, the fourth class has 26 students, the fifth class has 22 students, and the sixth class has 27 students. What is the median number of students in a seventh-grade math class at Sheepsend Middle School?

44. A car lot has 12 sports cars, eight sport utility vehicles, three convertibles, six pickup trucks, and six mid-size cars. If Tony bought a car today, what is the likelihood that he bought either a sport utility vehicle or a pickup truck?

45. A display rack contains ten postcards of New York, 20 postcards of Paris, eight postcards of London, four postcards of Rome, and three postcards of Tokyo. If Si buys two postcards, what is the probability that he bought a postcard of New York and a postcard of London?

46. If a polygon has eight sides, what is the sum of its interior angles?

47. The height of a triangle is 1.2 yards. If the area of the triangle is 1.98 square yards, what is the base of the triangle?

48. The hypotenuse of triangle *ABC* is 65 inches. If one leg of the triangle is 25 inches, what is the length of the other leg?

49. If the area of a circle is 169π square feet, what is the circumference of the circle?

50. The volume of a sphere is 288π cubic centimeters. What is the surface area of the sphere?

ANSWERS

1. *Read the entire word problem.*

   Identify the question being asked.

   We are looking for the difference between two numbers.

   Underline the keywords and words that indicate formulas.

   The keyword *difference* signals subtraction.

   Cross out extra information and translate words into numbers.

   There is no extra information in this problem.

   List the possible operations.

   We need to subtract 37 from 56.

   Write number sentences for each operation.

   56 – 37

   Solve the number sentences and decide which answer is reasonable.

   56 – 37 = 19

   Check your work.

   Since we used subtraction to solve this problem, we can use addition to check the answer. 19 + 37 = 56, so this answer is correct. For more on this concept, see Chapter 1.

2. *Read the entire word problem.*

   We are given the number of times a press can print a newspaper in an hour.

   Identify the question being asked.

   We are looking for the number of newspapers that six presses can print in an hour.

   Underline the keywords and words that indicate formulas.

   The keyword *each* can signal multiplication or division.

   Cross out extra information and translate words into numbers.

   There is no extra information in this problem.

   List the possible operations.

   We need to either divide 250 by 6 or multiply 250 by 6. Since we are given the number of newspapers printed by one press, and we are looking for the number printed by six presses, we must multiply.

   Write number sentences for each operation.

   (250)(6)

   Solve the number sentences and decide which answer is reasonable.

$(250)(6) = 1,500$ newspapers

Check your work.

Since we used multiplication to solve this problem, we can use division to check the answer. The number of newspapers divided by the number of newspapers printed by each press should give us the number of presses: $\frac{1,500}{250} = 6$ presses. For more on this concept, see Chapter 1.

3. *Read the entire word problem.*

We are given the number of rooms and the number of floors in a hotel.

Identify the question being asked.

We are looking for the number of rooms on each floor of the hotel.

Underline the keywords and words that indicate formulas.

The keyword *each* can signal multiplication or division.

Cross out extra information and translate words into numbers.

There is no extra information in this problem.

List the possible operations.

We need to either divide 360 by 24 or multiply 360 by 24. Since we are given the total number of rooms on 24 floors and we are looking for the number of rooms on one floor, we must divide.

Write number sentences for each operation.

$\frac{360}{15}$

Solve the number sentences and decide which answer is reasonable.

$\frac{360}{15} = 24$ rooms

Check your work.

Since we used division to solve this problem, we can use multiplication to check the answer. The number of rooms on each floor multiplied by the number of floors should give us the number of rooms in the hotel: $(24)(15) = 360$ rooms. For more on this concept, see Chapter 1.

4. *Read the entire word problem.*

We are given the number of dresses Terra has and the number of dresses Souyma has.

Identify the question being asked.

We are looking for the number of dresses the two have altogether.

Underline the keywords and words that indicate formulas.

The keyword *altogether* signals addition.

Cross out extra information and translate words into numbers.

There is no extra information in this problem.

List the possible operations.

We need to add the number of dresses Terra has to the number of dresses Souyma has.

Write number sentences for each operation.

18 + 15

Solve the number sentences and decide which answer is reasonable.

18 + 15 = 33 dresses

Check your work.

Since we used addition to solve this problem, we can use subtraction to check the answer. The total number of dresses minus the number of dresses Terra has should equal the number of dresses Souyma has: 33 − 18 = 15 dresses. For more on this concept, see Chapter 1.

5. *Read the entire word problem.*

We are given the number of times a hummingbird beats its wings in a second.

Identify the question being asked.

We are looking for the number of times it beats its wings in a minute.

Underline the keywords and words that indicate formulas.

The keyword *per* can signal multiplication or division.

Cross out extra information and translate words into numbers.

There is no extra information in this problem. There are 60 seconds in a minute, so this question is asking us how many times a hummingbird beats its wings in 60 seconds.

List the possible operations.

Since we are given the number of times a hummingbird beats its wings in one second and we are looking for the number of times it beats its wings in 60 seconds, we must multiply.

Write number sentences for each operation.

(53)(60)

Solve the number sentences and decide which answer is reasonable.

(53)(60) = 3,180 beats

Check your work.

Since we used multiplication to solve this problem, we can use division to check the answer. The total number of beats in a minute divided by the number of seconds in a minute should give us back the number of beats per second: $\frac{3,180}{60}$ = 53 beats. For more on this concept, see Chapter 2.

6. *Read the entire word problem.*

We are given the number of milliliters of iodine and the number of milliliters of water needed for a science experiment.

Identify the question being asked.

We are looking for the number of milliliters of iodine needed for 15 experiments.

Underline the keywords and words that indicate formulas.

There are no keywords in this problem.

Cross out extra information and translate words into numbers.

We do not need the number of milliliters of water used in each experiment, so that information can be crossed out.

List the possible operations.

Since we are given the number of milliliters of iodine used in one experiment and we are looking for the number of milliliters used in 15 experiments, we must multiply.

Write number sentences for each operation.

(5)(15)

Solve the number sentences and decide which answer is reasonable.

(5)(15) = 75 milliliters

Check your work.

Since we used multiplication to solve this problem, we can use division to check the answer. The total number of milliliters divided by the number of experiments should equal the number of milliliters of iodine needed for one experiment: $\frac{75}{15}$ = 5 milliliters. For more on this concept, see Chapter 2.

7. *Read the entire word problem.*

We are given the cost of skate rental, the current number of pairs of skates, and the number of skates that are ordered.

Identify the question being asked.

We are looking for the total number of skates.

Underline the keywords and words that indicate formulas.

The keyword *total* signals addition.

Cross out extra information and translate words into numbers.

We do not need the cost of skate rental, so that information can be crossed out.

List the possible operations.

We are given the current number of pairs of skates and the number of pairs of skates ordered, so we must add to find the total.

Write number sentences for each operation.

85 + 35

Solve the number sentences and decide which answer is reasonable.

85 + 35 = 120 pairs of skates

Check your work.

Since we used addition to solve this problem, we can use subtraction to check the answer. The total number of skates minus the number of skates ordered should equal the original number of skates: 120 − 35 = 85 pairs of skates. For more on this concept, see Chapter 2.

8. The following diagram shows 38 people waiting outside a theater:

17 of them are let in, so cross out 17 of them:

We can see now that there are 21 people still waiting outside. For more on this concept, see Chapter 3.

9. The following diagram shows 18 people at a party:

Six more people arrive, so add six guests to the diagram:

We can see now that there are 24 people at the party. For more on this concept, see Chapter 3.

10. We can enter these names into a table to find the order of their birthdays. Svetlana's birthday comes before Veronica's and after Cara's, so place Cara's name above Svetlana's name and Veronica's name:

| Cara |
| Svetlana |
| Veronica |

Cara's birthday is after Rachel's, so we can place Rachel at the top of the table:

| Rachel |
| Cara |
| Svetlana |
| Veronica |

We don't know where to place Susie's birthday, but we know it is after Cara's, which means it can be no higher than third. Rachel's party must be first. For more on this concept, see Chapter 3.

11. Build a table with columns for the number of booths, the number of tables, the total people from booths, the total people from tables, and the overall total. The overall total is equal to the number of booths multiplied by 4 plus the number of tables multiplied by 5. The total number of booths and tables should always equal 12:

Number of Booths	Number of Tables	Total from Booths	Total from Tables	Total
12	0	48	0	48
11	1	44	5	49
10	2	40	10	50
9	3	36	15	51
8	4	32	20	52
7	5	28	25	53

If there are seven booths and five tables, then there are 53 people seated in the restaurant. For more on this concept, see Chapter 3.

12. Draw a Venn diagram. We must draw it as shown here, so that there is an area where just the water park and amusement park overlap, an area where just the water park and the kiddie park overlap, an area where just the amusement park and the kiddie park overlap, and an area where all three overlap. We are given the number of people who visit only one park, so we can label those areas of the diagram. The unknown areas, the areas that represent people who visited multiple parks, are labeled with variables:

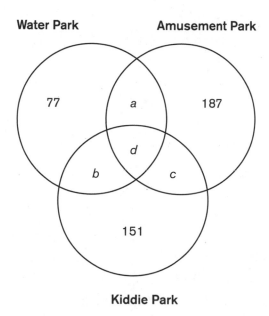

If 620 visited the park, and $77 + 187 + 151 = 415$ of them visited only one park, then $620 - 415 = 205$ visited more than one park, which means that $a + b + c + d = 205$. We know that 204 people visited the water park, which means that $77 + a + b + d = 204$. We also know that 370 people visited the amusement park, which means that $187 + a + d + c = 370$. In the same way, 295 people visited the kiddie park, which means that $151 + b + d + c = 295$. We can combine these equations to find the values of a, b, c, and d. If $77 + a + b + d = 204$, then $a + b + d = 127$. We know that $a + b + c + d = 205$. Subtract the first equation from the second:

$$
\begin{aligned}
a + b + c + d &= 205 \\
-(a + b + d &= 127) \\
\hline
c &= 78
\end{aligned}
$$

There are 78 people who visited the kiddie park and the amusement park. $187 + a + d + c = 370$, so $a + d + c = 183$. Subtract this equation from $a + b + c + d = 205$:

$$
\begin{aligned}
a + b + c + d &= 205 \\
-(a + d + c &= 183) \\
\hline
b &= 22
\end{aligned}
$$

There are 22 people who visited the water park and the kiddie park. Now that we know how many people visited the water park and the kiddie park (22) and how many people visited the kiddie park and the amusement park (78), we can find the number of people who visited all three parks, since we know 295 people visited the kiddie park: $151 + 22 + 78 + d = 295$, $251 + d = 295$, $d = 44$. There are 44 people who visited all three parks. For more on this concept, see Chapter 3.

13. *Read the entire word problem.*

Identify the question being asked.

We are looking for ten fewer than twelve.

Underline the keywords and words that indicate formulas.

The keywords *fewer than* signal subtraction.

Cross out extra information and translate words into numbers.

Rewrite *ten* as "10" and *twelve* as "12."

List the possible operations.

The keywords *fewer than* are a backward phrase; 12 is the number from which we are subtracting. We must find $12 - 10$.

Write number sentences for each operation.

$12 - 10$

Solve the number sentences and decide which answer is reasonable.

12 – 10 = 2

Check your work.

Since we used subtraction to solve this problem, we can use addition to check the answer. The difference, 2, plus the subtrahend, 10, should equal the minuend, 12: 2 + 10 = 12. For more on this concept, see Chapter 4.

14. *Read the entire word problem.*

We are given the number of fish Alimi catches and how many more fish he catches than Larry.

Identify the question being asked.

We are looking for the number of fish Larry catches.

Underline the keywords and words that indicate formulas.

The keywords *more than* usually signal addition, but in this problem, we already have the number of fish Alimi catches. We are looking for the number of fish Larry catches, which is less than the number of fish Alimi catches.

Cross out extra information and translate words into numbers.

There is no extra information in this problem.

List the possible operations.

Since Alimi catches more fish than Larry, we must subtract the difference in the number of fish each catches from Alimi's total to find how many fish Larry catches.

Write number sentences for each operation.

23 – 12

Solve the number sentences and decide which answer is reasonable.

23 – 12 = 11 fish

Check your work.

Since we used subtraction to solve this problem, we can use addition to check the answer. Since Alimi catches 12 more fish than Larry, 12 plus the number of fish Larry catches should equal the number of fish Alimi catches: 12 + 11 = 23 fish. For more on this concept, see Chapter 4.

15. *Read the entire word problem.*

We are given the number of books Hector buys at the convention and the number of books he has now.

Identify the question being asked.

We are looking for the number of books he had before the convention.

Underline the keywords and words that indicate formulas.

The keyword *adds* usually signals addition, but in this problem, we already have the number of books Hector has in his collection now. We are looking for the number of books he had before he added 23 to his collection.

Cross out extra information and translate words into numbers.

There is no extra information in this problem.

List the possible operations.

Since we already have the total number of books after Hector added 23 books, we must "undo" that addition and subtract to find how many books he had originally.

Write number sentences for each operation.

$289 - 23$

Solve the number sentences and decide which answer is reasonable.

$289 - 23 = 266$ books

Check your work.

Since we used subtraction to solve this problem, we can use addition to check the answer. The number of books Hector had before the convention plus the number of books he bought at the convention should equal the number of books he has now: $266 + 23 = 289$ books. For more on this concept, see Chapter 4.

16. *Read the entire word problem.*

We are given the number of ounces in the fountain, the number of ounces in each cup, and the number of cups.

Identify the question being asked.

We are looking for the number of ounces that are left in the fountain.

Underline the keywords and words that indicate formulas.

The keyword *left* signals subtraction, but before we can subtract, we must determine how many ounces to subtract.

Cross out extra information and translate words into numbers.

There is no extra information in this problem.

List the possible operations.

One cup of water contains 8 ounces, and there are 45 cups. We must multiply to find the total number of ounces removed from the fountain. Once we have that value, we can subtract it from the original number of ounces in the fountain.

Write number sentences for each operation.

Find the number of ounces removed from the fountain:

(45)(8)

Solve the number sentences and decide which answer is reasonable.

(45)(8) = 360 ounces

Write number sentences for each operation.

Now subtract that amount from the original number of ounces in the fountain:

640 – 360

Solve the number sentences and decide which answer is reasonable.

640 – 360 = 280 ounces

Check your work.

Since we used multiplication and subtraction to solve this problem, we can use addition and division to check the answer. Divide the number of ounces in 45 cups, 360, by 45 to find the amount of water in one cup: $\frac{360}{45}$ = 8 ounces. Add the amount of water removed from the fountain, 360 ounces, to the amount of water remaining in the fountain, 280 ounces, to check that it is equal to the original volume of the fountain: 360 + 280 = 640 ounces. For more on this concept, see Chapter 5.

17. *Read the entire word problem.*

We are given the number of pounds in a knapsack, the number of pounds in a duffel bag, and the numbers of knapsacks and duffel bags. *Identify the question being asked.*

We are looking for the total weight of the baggage.

Underline the keywords and words that indicate formulas.

The keyword *each* can signal multiplication or division, and the keyword appears twice in the problem. The keyword *add* signals addition.

Cross out extra information and translate words into numbers.

There is no extra information in this problem.

List the possible operations.

If one knapsack weighs 12 pounds, then we must multiply to find the weight of 12 knapsacks. In the same way, if one duffel bag weighs 20 pounds, then we must multiply to find the weight of five duffel bags. Once we have those weights, we can add them to find the total weight of the baggage.

Write number sentences for each operation.

Find the weight of the knapsacks:

(12)(12)

Solve the number sentences and decide which answer is reasonable.

(12)(12) = 144 pounds

Write number sentences for each operation.

Next, find the weight of the duffel bags:

(20)(5)

Solve the number sentences and decide which answer is reasonable.

(20)(5) = 100 pounds

Write number sentences for each operation.

Finally, add the weight of the knapsacks to the weight of the duffel bags to find the total weight:

144 + 100

Solve the number sentences and decide which answer is reasonable.

144 + 100 = 244 pounds

Check your work.

Retrace your steps. Divide the total weight of the knapsacks, 144 pounds, by the number of knapsacks: $\frac{144}{12}$ = 12 pounds per knapsack, which is what was given. Divide the total weight of the duffel bags, 100 pounds, by the number of duffel bags: $\frac{100}{5}$ = 20 pounds per bag, which is what was given. For more on this concept, see Chapter 5.

18. *Read the entire word problem.*

We are given the number of people one bus holds, the number of trips made by one bus, and the number of buses.

Identify the question being asked.

We are looking for the total number of people served in five days.

Underline the keywords and words that indicate formulas.

The keyword *per* can signal multiplication or division.

Cross out extra information and translate words into numbers.

There is no extra information in this problem.

List the possible operations.

If one bus holds 22 people, then to find the number of people served by eight trips, we must multiply. Once we have that product, we can multiply it by nine, the number of buses the tour company has. This will give us the number of people served in one day. To find the number of people served in five days, we will have to multiply again.

Write number sentences for each operation.

First, find the number of people served by eight trips:

(22)(8)

Solve the number sentences and decide which answer is reasonable.

(22)(8) = 176 people

Write number sentences for each operation.

Next, find the number of people served by nine buses:

(176)(9)

Solve the number sentences and decide which answer is reasonable.

(176)(9) = 1,584 people

Write number sentences for each operation.

Since 1,584 people are served in one day, multiply that by 5 to find the number of people served in five days:

(1,584)(5)

Solve the number sentences and decide which answer is reasonable.

(1,584)(5) = 7,920 people

Check your work.

We can use division to check our answer. Divide the number of people served in five days by 5 to find the number of people served in one day: $\frac{7,920}{5} = 1,584$. Divide by 9 the number of people served in one day by nine buses to find the number of people served by one bus: $\frac{1,584}{9} = 176$. Divide by 8 the number of people served by 8 trips to find the number of people served by one trip: $\frac{176}{8} = 22$, the given number of people the tour bus holds. For more on this concept, see Chapter 5.

19. *Read the entire word problem.*

Identify the question being asked.

We are looking for the sum of six fewer than eleven and nineteen.

Underline the keywords and words that indicate formulas.

The keyword *sum* signals addition, and the keyword phrase *fewer than* signals subtraction.

Cross out extra information and translate words into numbers.

Rewrite *six* as "6," *eleven* as "11," *nineteen* as "19."

List the possible operations.

We are looking for the sum of 19 and a number. That number is 6 fewer than 11, so we must subtract 6 from 11 before adding.

Write number sentences for each operation.

11 – 6

Solve the number sentences and decide which answer is reasonable.

11 – 6 = 5

Write number sentences for each operation.

Now, find the sum of 5 and 19:

5 + 19

Solve the number sentences and decide which answer is reasonable.

5 + 19 = 24

Check your work.

We can use subtraction to check our answer. Subtract 19 from 24, and you should be left with a number that is 6 fewer than 11. 24 – 19 = 5, as is 11 – 6. For more on this concept, see Chapter 5.

20. *Read the entire word problem.*

Identify the question being asked.

We are looking for the difference of two numbers, the first of which is the product of eight and fourteen, and the second of which is the sum of twelve and a number that is four more than nine.

Underline the keywords and words that indicate formulas.

The keyword *product* signals multiplication, the keyword *minus* signals subtraction, the keyword *sum* signals addition, and the keyword phrase *more than* signals addition.

Cross out extra information and translate words into numbers.

Rewrite *eight* as "8," *fourteen* as "14," *twelve* as "12," *four* as "4," and *nine* as "9."

List the possible operations.

We are looking for the difference between the product of 8 and 14 and a sum, but before we can find the difference, we must find the sum of 12 and 4 more than 9.

Write number sentences for each operation.

To find 4 more than 9, add 4 to 9:

4 + 9

Solve the number sentences and decide which answer is reasonable.

4 + 9 = 13

Write number sentences for each operation.

Now, find the sum of 12 and 13:

12 + 13

Solve the number sentences and decide which answer is reasonable.

12 + 13 = 25

Write number sentences for each operation.

Next, find the product of 8 and 14:

(8)(14)

Solve the number sentences and decide which answer is reasonable.

(8)(14) = 112

Write number sentences for each operation.

Finally, subtract from that product the sum of 12 and 4 more than 9:

112 – 25

Solve the number sentences and decide which answer is reasonable.

112 – 25 = 87

Check your work.

Retrace your steps. Add 25 to 87: 25 + 87 = 112. Divide 112 by 14: $\frac{112}{14}$ = 8, one of the given numbers. Subtract 12 from 25: 25 – 12 = 13. Thirteen is 4 more than 9, two of the other given numbers. For more on this concept, see Chapter 5.

21. Write *seven* as "7," *a number* as "*x*," and *thirteen* as "13." *Product* signals multiplication, so the product of *x* and 13 is $13x$. *More than* signals addition, so 7 more than $13x$ is $13x + 7$. For more on this concept, see Chapter 6.

22. Write *half* as "$\frac{1}{2}$," *a number* as "*x*," and *fifteen* as "15." The square of *x* is x^2. Half the square of a number is $\frac{1}{2}x^2$. *Times* signals multiplication, so 15 times *x* is $15x$. *Minus* signals subtraction, so $\frac{1}{2}x^2$ minus $15x$ is $\frac{1}{2}x^2 - 15x$. For more on this concept, see Chapter 6.

23. Write *three* as "3," *a number* as "*x*," write *eleven* as "11," and write *five* as "5." Replace *is equal to* with the equals sign: the sum of 3 times *x* and 11 = 5. *Sum* signals addition and *times* signals multiplication: The sum of 3 times *x* and 11 = 5 becomes $3x + 11 = 5$. Subtract 11 from both sides of the equation, and then divide both sides by 3: $3x + 11 = 5$, $3x = -6$, $x = -2$. For more on this concept, see Chapter 6.

24. Write *one* as "1," write *nine* as "9," write *a number* as "*x*," write *five* as "5," and write *eight* as "8." Replace *is* with the equals sign: 1 fewer than 9 times *x* = 5 more than 8 times *x*. *Fewer than* signals subtraction, *times* signals multiplication, and *more than* signals addition: 1 fewer than 9 times *x* = 5 more than 8 times *x* becomes $9x - 1 = 8x + 5$. Subtract $8x$ from both sides of the equation and add 1 to both sides: $9x - 1 = 8x + 5$, $x = 6$. For more on this concept, see Chapter 6.

25. Write *negative six* as "–6," write *a number* as "*x*," write *four* as "4," write *twelve* as "12," and write *two* as "2." Replace *is less than* with the less than symbol: –6 times *x* plus 4 < 12 fewer than 2 multiplied by *x*. *Times* signals multiplication, *plus* signals addition, *fewer than* signals subtraction, and *multiplied by* signals multiplication: –6 times *x* plus 4 < 12 fewer than 2 multiplied by *x* becomes $-6x + 4 < 2x - 12$. Subtract $2x$ and –4 from both sides of the inequality. $-6x + 4 < 2x - 2$, $-8x < -16$. Divide both sides of the inequality by –8 and reverse the direction of the inequality symbol: $-8x < -16$, $x > 2$. For more on this concept, see Chapter 6.

26. *Read the entire word problem.*

We are given the fraction of Marco's gas tank that was full and the fraction that he uses.

Identify the question being asked.

We are looking for the fraction of his gas tank that is full now.

Underline the keywords and words that indicate formulas.

There are no keywords in this problem.

Cross out extra information and translate words into numbers.

There is no extra information in this problem.

List the possible operations.

Since Marco uses $\frac{1}{4}$ of the tank, he has $\frac{1}{4}$ less, which means that we must subtract $\frac{1}{4}$ from $\frac{5}{8}$.

Write number sentences for each operation.

Before we can subtract, we must have common denominators. The least common multiple of 4 and 8 is 8, so convert $\frac{1}{4}$ to eighths: $\frac{1}{4} = \frac{2}{8}$.

$\frac{5}{8} - \frac{2}{8}$

Solve the number sentences and decide which answer is reasonable.

$\frac{5}{8} - \frac{2}{8} = \frac{3}{8}$

$\frac{3}{8}$ of Marco's gas tank is full now.

Check your work.

Add the fraction of the tank that remains to the fraction of the tank Marco used over the weekend. This sum should equal the fraction of the tank that was originally full. $\frac{3}{8} + \frac{1}{4} = \frac{3}{8} + \frac{2}{8} = \frac{5}{8}$. For more on this concept, see Chapter 7.

27. *Read the entire word problem.*

We are given the number of pounds of chocolate Gino buys, the fraction of pounds that is white chocolate, and the number of friends with whom Gino shares.

Identify the question being asked.

We are looking for the number of pounds of chocolate each friend receives.

Underline the keywords and words that indicate formulas.

The keyword *shares* signals division.

Cross out extra information and translate words into numbers.

There is no extra information in this problem.

List the possible operations.

First, we must find what fraction of the chocolate is dark chocolate. Then, we must multiply that by 2, the total number of pounds of chocolate Gino bought. Finally, we will divide that quantity by 5, since Gino shared the dark chocolate with five friends.

Write number sentences for each operation.

If $\frac{3}{10}$ of the chocolate was white chocolate and the rest was dark chocolate, we must subtract $\frac{3}{10}$ from 1 to find the fraction of dark chocolate. Write 1 as a fraction with a denominator of 10:

$\frac{10}{10} - \frac{3}{10}$

Solve the number sentences and decide which answer is reasonable.

$\frac{10}{10} - \frac{3}{10} = \frac{7}{10}$

$\frac{7}{10}$ of the chocolate is dark chocolate.

Write number sentences for each operation.

$\frac{7}{10}$ of the 2 pounds of chocolate Gino bought is dark chocolate. Multiply $\frac{7}{10}$ by 2 to find the number of pounds of dark chocolate:

$(2)(\frac{1}{10})$

Solve the number sentences and decide which answer is reasonable.

$(2)(\frac{7}{10}) = \frac{14}{10}$

Gino has $\frac{14}{10}$ pounds of dark chocolate.

Write number sentences for each operation.

Gino shares $\frac{14}{10}$ pounds of dark chocolate with five friends. Divide $\frac{14}{10}$ by 5:

$$\frac{\left(\frac{14}{10}\right)}{5}$$

Solve the number sentences and decide which answer is reasonable.

$$\frac{\left(\frac{14}{10}\right)}{5} = \frac{14}{50} = \frac{7}{25}$$

Each friend receives $\frac{7}{25}$ pounds of dark chocolate.

Check your work.

Multiply the number of pounds of chocolate each friend receives by the number of friends: $\left(\frac{7}{25}\right)(5) = \frac{7}{5}$ pounds. Since Gino bought 2 pounds of chocolate, and $\frac{7}{10}$ of it was dark chocolate, $\frac{7}{5}$ divided by $\frac{7}{10}$ should equal 2 pounds: $\frac{\left(\frac{7}{5}\right)}{\left(\frac{7}{10}\right)} = \left(\frac{7}{5}\right)\left(\frac{10}{7}\right) = 2$ pounds. For more on this concept, see Chapter 7.

28. *Read the entire word problem.*

We are given the number of ounces of water the faucet loses per hour.

Identify the question being asked.

We are looking for the number of ounces of water it loses in a day.

Underline the keywords and words that indicate formulas.

The keyword *per* can signal multiplication or division.

Cross out extra information and translate words into numbers.

There is no extra information in this problem. We are told how many ounces of water are lost each hour, and we are looking for how many ounces are lost in a day, so we must convert one day to 24 hours.

List the possible operations.

Multiply the number of ounces of water lost in one hour by 24 hours to find the number of ounces lost in a day.

Write number sentences for each operation.

(0.06)(24)

Solve the number sentences and decide which answer is reasonable.

(0.06)(24) = 1.44 ounces

The faucet loses 1.44 ounces in a day.

Check your work.

The number of ounces lost in a day divided by 24 should equal the number of ounces lost in an hour: $\frac{1.44}{24} = 0.06$ ounces. For more on this concept, see Chapter 7.

29. *Read the entire word problem.*

We are given the size of Christian's hard drive, the amount of data he had saved, and the amount of data he deleted.

Identify the question being asked.

We are looking for the amount of free space on his hard drive.

Underline the keywords and words that indicate formulas.

There are no keywords in this problem.

Cross out extra information and translate words into numbers.

There is no extra information in this problem.

List the possible operations.

If Christian deletes data from his hard drive, then we must subtract that amount from the amount of data he had saved. Once we have the new amount of data he has saved, we can subtract that from the size of the hard drive to find how many gigabytes are free.

Write number sentences for each operation.

First, find how many gigabytes are used. Subtract the amount Christian deleted from the amount he had saved:

26.754 − 8.24

Solve the number sentences and decide which answer is reasonable.

26.754 − 8.24 = 18.514 gigabytes

Write number sentences for each operation.

Now, subtract the amount of space used from the size of the drive to find the number of gigabytes that are free:

50 − 18.514

Solve the number sentences and decide which answer is reasonable.

50 − 18.514 = 31.486 gigabytes

Check your work.

Subtract the amount of free space from the total size of the drive. This will give you the amount of used space. Add to that the number of gigabytes Christian deleted, and this should give you the original number of gigabytes saved on the drive: 50 − 31.486 = 18.514, 18.514 + 8.24 = 26.754 gigabytes. For more on this concept, see Chapter 7.

30. *Read the entire word problem.*

We are given the number of appointments Dr. Wilcox can schedule in a week and the number of appointments he has scheduled.

Identify the question being asked.

We are looking for the percent of his schedule that is booked.

Underline the keywords.

There are no keywords in this problem, but the problem does contain the word *percent*, so we will likely have to use a percent formula.

Cross out extra information and translate words into numbers.

There is no extra information in this problem.

List the possible operations.

We need to find what percent is 21 of 30. We can find what percent one number is of a second number by dividing the first number by the second number. Divide 21 by 30 and express the answer as a percent.

Write number sentences for each operation.

$\frac{21}{30}$

Solve the number sentences and decide which answer is reasonable.

$\frac{21}{30} = 0.70 = 70\%$

70% of Dr. Wilcox's schedule is booked.

Check your work.

Since we used division to find our answer, we can check our work with multiplication. Multiply the percent of Dr. Wilcox's schedule that is booked by the total number of appointments he could have. This should give us back the number of appointments he has booked: $(0.70)(30) = 21$ appointments. For more on this concept, see Chapter 7.

31. *Read the entire word problem.*

We are given the original price of a photo, the percent that it increased, and the percent that it then decreased.

Identify the question being asked.

We are looking for the final price of the photo.

Underline the keywords.

The keyword *increased* usually signals addition, but the problem does contain the percent symbol twice, so we will likely have to use at least one percent formula.

Cross out extra information and translate words into numbers.

There is no extra information in this problem.

List the possible operations.

First, we need to find how much the price of the photo increased. We will multiply the original price by 15%, or 0.15, and add that increase to the original price. Then, we will find 25% of that price, and subtract that amount to give us the final price.

Write number sentences for each operation.

Multiply the original price, $120, by 15%, or 0.15:

($120)(0.15)

Solve the number sentences and decide which answer is reasonable.

($120)(0.15) = $18

The price of the photo increased by $18.

Write number sentences for each operation.

Add $18 to the original price:

$120 + $18

Solve the number sentences and decide which answer is reasonable.

$120 + $18 = $138

The price of the photo became $138.

Write number sentences for each operation.

Next, find 25% of $138. Multiply $138 by 25%, or 0.25:

($138)(0.25)

Solve the number sentences and decide which answer is reasonable.

($138)(0.25) = $34.50

The price of the photo decreased by $34.50.

Write number sentences for each operation.

Find the final price by subtracting $34.50 from $138:

$138 – $34.50

Solve the number sentences and decide which answer is reasonable.

$138 – $34.50 = $103.50

The price of the photo now is $103.50.

Check your work.

There is no easy way to check a problem like this. Review your calculations to ensure that this answer is correct. For more on this concept, see Chapter 7.

32. *Read the entire word problem.*

We are given a temperature in degrees Fahrenheit.

Identify the question being asked.

We are looking for a temperature in degrees Celsius.

Underline the keywords and words that indicate formulas.

The words *Celsius* and *Fahrenheit* tell us that we need to use a temperature conversion formula.

Cross out extra information and translate words into numbers.

There is no extra information in this problem.

List the possible operations.

We need to use the formula for converting temperature in degrees Fahrenheit to degrees Celsius: $C = \frac{5}{9}(F - 32)$.

Write number sentences for each operation.

Substitute the temperature in degrees Fahrenheit into the formula:

$C = \frac{5}{9}(77 - 32)$

Solve the number sentences and decide which answer is reasonable.

$C = \frac{5}{9}(77 - 32) = \frac{5}{9}(45) = 25°$ Celsius.

Check your work.

To check our answer, we can use the formula for converting Celsius to Fahrenheit to get back the temperature in degrees Fahrenheit. $F = \frac{9}{5}C + 32$, $F = \frac{9}{5}(25) + 32 = 45 + 32 = 77°$ Fahrenheit. For more on this concept, see Chapter 7.

33. *Read the entire word problem.*

We are given the principal, rate, and time.

Identify the question being asked.

We are looking for the interest earned by that principal at that rate over that time.

Underline the keywords and words that indicate formulas.

The keyword *per* can signal multiplication or division. The words *interest* and *rate* tell us that we need to use the formula for interest.

Cross out extra information and translate words into numbers.

There is no extra information in this problem.

List the possible operations.

To find interest earned given the principal, rate, and time, we must multiply those three values. However, the interest rate is 4.2% per year, and our time is given in months. We must first convert three months to a number of years.

Write number sentences for each operation.

There are 12 months in a year, so we can convert the number of months to a number of years by dividing by 12:

$\frac{3}{12}$

Solve the number sentences and decide which answer is reasonable.

$\frac{3}{12} = 0.25$

Janine had her money in the bank for 0.25 years. We are now ready to use the interest formula, $I = prt$, where I, interest, is equal to the product of p, principal, r, rate, and t, time.

Write number sentences for each operation.

Convert 4.2% to a decimal. Our number sentence is the interest formula, with 8,200 substituted for p, 0.042 substituted for r, and 0.25 substituted for t:

$I = (8,200)(0.042)(0.25)$

Solve the number sentences and decide which answer is reasonable.

$I = (8,200)(0.042)(0.25) = \$86.10$

Check your work.

Since $I = prt$, we can check our interest answer by dividing the interest by the rate and time to see if that quotient equals the given principal, $8,200: \frac{\$86.10}{(.042)(0.25)} = \frac{\$86.10}{0.0105} = \$8,200$. For more on this concept, see Chapter 7.

34. *Read the entire word problem.*

We are given the rate at which the roller coaster travels and the time for which it travels.

Identify the question being asked.

We are looking for the distance it travels.

Underline the keywords and words that indicate formulas.

The keyword *per* can signal multiplication or division.

Cross out extra information and translate words into numbers.

There is no extra information in this problem.

List the possible operations.

To find the distance an object travels, we must multiply the rate by the time. However, the rate is given in miles per hour and the time is given in seconds. Both measurements must use the same unit of time. We will convert the number of seconds to hours. There are 60 seconds in a minute, and 60 minutes in an hour, which means that there are $(60)(60) = 3,600$ seconds in an hour. Divide the number of seconds by 3,600 to represent the time in hours.

Write number sentences for each operation.

$\frac{4}{3,600} = \frac{1}{900}$

The roller coaster traveled for $\frac{1}{900}$ of an hour. The keyword *per* signals multiplication, since the rate gives us the distance traveled in an hour, and we are looking for the distance traveled in $\frac{1}{900}$ of an hour.

Write number sentences for each operation.

Our number sentence is the distance formula, with 65 substituted for r and $\frac{1}{900}$ substituted for t: $D = (65)(\frac{1}{900})$

Solve the number sentences and decide which answer is reasonable.

$(65)(\frac{1}{900}) = \frac{65}{900} = 0.072$ miles, to the nearest thousandth.

Check your work.

Since $D = rt$, we can check our distance answer by dividing the distance by the time to see if that quotient equals the given rate, 65 miles per hour: $\frac{0.072}{(\frac{1}{900})} = 64.8$, or about 65 miles per hour. For more on this concept, see Chapter 8.

35. Make a table with three rows and three columns. Since we are looking for how much salt will be added, our table will be about the percent and quantity of salt. The original solution is 5%, or 0.05, salt. It is 20 ounces in total, which means that 0.05×20 of it is salt. We are adding an unknown quantity of salt, 100% of which is salt. Our final solution will be 50% salt:

	Percent concentration	Total quantity (oz.)	Salt quantity (oz.)
Original solution	0.05	20	0.05×20
Added	1	x	$1 \times x$
Final solution	0.50		

The total quantity of the final solution will be $20 + x$, and 0.50 of it will be salt:

	Percent concentration	Total quantity (oz.)	Salt quantity (oz.)
Original solution	0.05	20	0.05×20
Added	1	x	$1 \times x$
Final solution	0.50	$20 + x$	$0.50(20 + x)$

The salt quantity is the product of the percent concentration and the total quantity, $0.50(20 + x)$, and also the sum of the salt quantity in the original solution and the amount added, 0.05×20 and $1 \times x$. Set these two expressions equal to each other and solve for x:

$0.50(20 + x) = (0.05 \times 20) + (1 \times x)$
$10 + 0.5x = 1 + x$
$9 = 0.5x$
$x = 18$ ounces

Ida must add 18 ounces of salt to make the solution 50% salt. For more on this concept, see Chapter 10.

36. Make a table with three rows and three columns. Since we are looking for how much water will be evaporated, our table will be about the percent and quantity of water. The solution is 15% alcohol and $100\% - 15\%$ $= 85\%$ water. We are looking for a final solution that is only $100\% - 75\%$ $= 25\%$ water:

	Percent concentration	Total quantity (L)	Water quantity (L)
Original solution	0.85	45	0.85×45
Removed	1	x	$1 \times x$
Final solution	0.25		

We are removing water, so the total quantity will be $45 - x$, not $45 + x$:

	Percent concentration	Total quantity (L)	Water quantity (L)
Original solution	0.85	45	0.85×45
Removed	1	x	$1 \times x$
Final solution	0.25	$45 - x$	$0.25(45 - x)$

The final water quantity is equal to the percent concentration of the final solution times the total quantity of the solution: $0.25(45 - x)$. The final quantity of water is also equal to the original quantity of water minus the amount of water removed: $(0.85 \times 45) - (1 \times x)$. Set these equations equal to each other and solve for x:

$0.25(45 - x) = (0.85 \times 45) - (1 \times x)$
$11.25 - 0.25x = 38.25 - x$

$0.75x = 27$
$x = 36$ liters

We must evaporate 36 liters of water to make the concentration of the solution 75% alcohol. For more on this concept, see Chapter 10.

37. *Read the entire word problem.*
We are given the number of necklaces and the number of earrings.
Identify the question being asked.
We are looking for the ratio of earrings to necklaces.
Underline the keywords and words that indicate formulas.
The word *ratio* usually means that we will be using a ratio to set up a proportion, but this problem simply asks us to find a ratio.
Cross out extra information and translate words into numbers.
There is no extra information in this problem.
List the possible operations.
A ratio is a relationship between two numbers. There are no number sentences to write: the ratio of earrings to necklaces is the number of earrings, followed by a colon, and then the number of necklaces: 30:12. We can reduce this ratio by dividing both numbers by their greatest common factor. The greatest common factor of 30 and 12 is 6, so 30:12 reduces to 5:2. For more on this concept, see Chapter 8.

38. *Read the entire word problem.*
We are given the number of left-handed players and right-handed players on a team.
Identify the question being asked.
We are looking for the ratio of left-handed players to the total number of players.
Underline the keywords and words that indicate formulas.
The word *ratio* usually means that we will be using a ratio to set up a proportion, but this problem simply asks us to find a ratio. The keyword *total* signals addition.
Cross out extra information and translate words into numbers.
There is no extra information in this problem.
List the possible operations.
A ratio is a relationship between two numbers. To find the ratio of left-handed players to the total number of players, we must first the total

number of players. Add the number of left-handed players and the number of right-handed players.

Write number sentences for each operation.

14 + 6

Solve the number sentences and decide which answer is reasonable.

14 + 6 = 20 players

The ratio of left-handed players to the total number of players is the number of left-handed players, followed by a colon, and then the total number of players: 14:20. We can reduce this ratio by dividing both numbers by their greatest common factor. The greatest common factor of 14 and 20 is 2, so 14:20 reduces to 7:10.

Check your work.

We can check our addition by using subtraction. The total number of players minus the number of left-handed players should equal the number of right-handed players: 20 − 14 = 6 players. For more on this concept, see Chapter 8.

39. *Read the entire word problem.*

We are given the ratio of cucumber slices to tomato slices and the number of cucumber slices altogether.

Identify the question being asked.

We are looking for the number of tomato slices.

Underline the keywords and words that indicate formulas.

The word *ratio* usually means that we will be using a ratio to set up a proportion.

Cross out extra information and translate words into numbers.

There is no extra information in this problem.

List the possible operations.

This problem contains two quantities, cucumber slices and tomato slices. We are given the value of one quantity, cucumber slices, and we are asked for the value of the other quantity, tomato slices. This is a part-to-part problem. We must set up a proportion that compares the ratio of cucumber slices to tomato slices to the ratio of actual cucumber slices to actual tomato slices.

Write number sentences for each operation.

The ratio of cucumber slices to tomato slices is 4:3, or $\frac{4}{3}$. Use x to represent the actual number of tomato slices, since that number is unknown. The ratio of actual cucumber slices to actual tomato slices is 12:*x*, or $\frac{12}{x}$. Set these fractions equal to each other:

$$\frac{4}{3} = \frac{12}{x}$$

Solve the number sentences and decide which answer is reasonable.

Cross multiply and solve for *x*: $4x = (12)(3)$, $4x = 36$, $x = 9$. If there are 12 cucumber slices in the salad, then there are nine tomato slices in the salad.

Check your work.

The ratio of cucumber slices to tomato slices is 4:3, so the ratio of actual cucumber slices to actual tomato slices should reduce to 4:3. The greatest common factor of 12 and 9 is 3, and 12:9 reduces to 4:3. For more on this concept, see Chapter 8.

40. *Read the entire word problem.*

We are given the ratio of first-class seats to business-class seats and the total number of seats altogether.

Identify the question being asked.

We are looking for the number of first-class seats.

Underline the keywords and words that indicate formulas.

The word *ratio* means that we will likely be using a ratio to set up a proportion. The word *total* is a signal that this is a part-to-whole ratio problem.

Cross out extra information and translate words into numbers.

There is no extra information in this problem.

List the possible operations.

This problem contains two quantities, first-class seats and business-class seats. We are given the total of the two quantities, and we are looking for the value of one quantity. This is a part-to-whole problem. Since we are looking for the number of first-class seats, we must set up a proportion that compares the ratio of first-class seats to total seats to the ratio of actual first-class seats to actual total seats.

Write number sentences for each operation.

The ratio of first-class seats to business-class seats is 2:25, which means that the ratio of first-class seats to total seats is 2:25 + 2, which is 2:27, or $\frac{2}{27}$. Use *x* to represent the number of first-class seats, since that number is unknown. There are 324 total seats, so the ratio of actual first-class seats to actual total seats is x:324, or $\frac{x}{324}$. Set these fractions equal to each other:

$$\frac{2}{27} = \frac{x}{324}$$

Solve the number sentences and decide which answer is reasonable.

Cross multiply and solve for x:

$27x = (2)(324)$, $27x = 648$, $x = 24$. If there are 324 total seats, then 24 of them are first-class seats.

Check your work.

The ratio of first-class seats to total seats is 2:27, so the ratio of actual first-class seats to total seats should reduce to 2:27. The greatest common factor of 24 and 324 is 12, and 24:324 reduces to 2:27. For more on this concept, see Chapter 8.

41. *Read the entire word problem.*

We are given the ratio of red tiles to blue tiles and the number of red tiles.

Identify the question being asked.

We are looking for the total number of tiles.

Underline the keywords and words that indicate formulas.

The word *ratio* means that we will likely be using a ratio to set up a proportion. The word *total* is a signal that this is a part-to-whole ratio problem.

Cross out extra information and translate words into numbers.

There is no extra information in this problem.

List the possible operations.

This problem contains two quantities, red tiles and blue tiles. We are given the value of one quantity, and we are looking for the total of the two quantities. This is a part-to-whole problem. Since we are looking for the total number of tiles, we must set up a proportion that compares the ratio of red tiles to the total number of tiles to the ratio of actual red tiles to the actual total number tiles.

Write number sentences for each operation.

The ratio of red tiles and blue tiles is 3:8, which means that the ratio of red tiles to the total number of tiles is 3:8 + 3, which is 3:11, or $\frac{3}{11}$. Use x to represent the number of total tiles, since that number is unknown. There are 1,692 red tiles, so the ratio of actual red tiles to actual total tiles is 1,692:x, or $\frac{1,692}{x}$. Set these fractions equal to each other:

$$\frac{3}{11} = \frac{1,692}{x}$$

Solve the number sentences and decide which answer is reasonable.

Cross multiply and solve for x:

$3x = (11)(1,692)$, $3x = 18,612$, $x = 6,204$. If there are 1,692 red tiles, then there are 6,204 total tiles.

Check your work.

The ratio of red tiles to the total number of tiles is 3:11, so the ratio of actual red tiles to the total number of actual tiles should reduce to 3:11. The greatest common factor of 1,692 and 6,204 is 564, and 1,692:6,204 reduces to 3:11. For more on this concept, see Chapter 8.

42. _Read the entire word problem._

We are given the number of visitors to a library.

Identify the question being asked.

We are looking for the range of visitors to the library.

Underline the keywords and words that indicate formulas.

The word _range_ actually appears in the problem, so we are told that we are looking for a range.

Cross out extra information and translate words into numbers.

The number of weeks, two, is not needed to solve this problem, so it can be crossed out.

List the possible operations.

The range is found by subtracting the smallest value from the largest value.

Write number sentences for each operation.

The smallest number in the set is 21 visitors and the largest number in the set is 77 visitors:

77 – 21

Solve the number sentences and decide which answer is reasonable.

77 – 21 = 56 visitors

Check your work.

Since we used subtraction to find our answer, we can use addition to check our answer. The smallest value plus the range should equal the largest value: 21 + 56 = 77, which is the largest value in the set. For more on this concept, see Chapter 9.

43. _Read the entire word problem._

We are given the number of students in six math classes.

Identify the question being asked.

We are looking for the median number of students in a math class.

Underline the keywords and words that indicate formulas.

The word *median* actually appears in the problem, so we are told that we are looking for a median.

Cross out extra information and translate words into numbers.

There is no extra information in this problem.

List the possible operations.

To find the median value, put all of the class sizes in order from smallest to greatest and select the middle value. Since there is an even number of classes, the median value will be the average of the third and fourth scores.

Write number sentences for each operation.

First, we must place the numbers of students in order from smallest to greatest:

22, 23, 24, 26, 27, 27

The middle values are 24 and 26. The average of these numbers is equal to their sum divided by 2, so first we must find the sum:

24 + 26

Solve the number sentences and decide which answer is reasonable.

24 + 26 = 50

Write number sentences for each operation.

Now, divide that sum by 2:

$\frac{50}{2}$

Solve the number sentences and decide which answer is reasonable.

$\frac{50}{2} = 25$

Check your work.

We can check that our median could be correct by comparing the number of values that are less than or equal to the median to the number of values that are greater than or equal to the median. These numbers should be equal. There are three values that are less than 25 (22, 23, 24) and three values that are greater than 25 (26, 27, 27), so 25 could be the median. For more on this concept, see Chapter 9.

44. *Read the entire word problem.*

We are told a car lot has 12 sports cars, eight sport utility vehicles, three convertibles, six pickup trucks, and six mid-size cars.

Identify the question being asked.

We are looking for the likelihood that Tony bought either a sport utility vehicle or a pickup truck.

Underline the keywords and words that indicate formulas.

The word *likelihood* tells us that we are looking for a probability. The word *or* tells us that we are looking for two probabilities, which we will likely have to add.

Cross out extra information and translate words into numbers.

There is no extra information in this problem.

List the possible operations.

The total number of possibilities is the total number of cars in the lot, so we must first find that total.

Write number sentences for each operation.

$12 + 8 + 3 + 6 + 6$

Solve the number sentences and decide which answer is reasonable.

$12 + 8 + 3 + 6 + 6 = 35$

Write number sentences for each operation.

Since there are eight sport utility vehicles, the probability of Tony buying one of those is $\frac{8}{35}$. There are six pickup trucks, so the probability of Tony buying one of those is $\frac{6}{35}$. To find the probability that he bought either a sport utility vehicle or a pickup truck, add the two probabilities:
$\frac{8}{35} + \frac{6}{35}$

Solve the number sentences and decide which answer is reasonable.

$\frac{8}{35} + \frac{6}{35} = \frac{14}{35}$, or $\frac{2}{5}$

The probability that Tony bought either a sport utility vehicle or a pickup truck is $\frac{2}{5}$. For more on this concept, see Chapter 9.

45. *Read the entire word problem.*

We are told a display rack contains ten postcards of New York, 20 postcards of Paris, eight postcards of London, four postcards of Rome, and three postcards of Tokyo.

Identify the question being asked.

We are looking for the probability that Si bought a postcard of New York and a postcard of London.

Underline the keywords and words that indicate formulas.

We are told that we are looking for a probability, and the keyword *and* tells us that we are looking for two probabilities, which we will likely have to multiply. Since Si buys two postcards, the denominator of our second probability will not be the same as the denominator of our first probability.

Cross out extra information and translate words into numbers.

There is no extra information in this problem.

List the possible operations.

The total number of possibilities is the total number of postcards, so we must first find that total.

Write number sentences for each operation.

10 + 20 + 8 + 4 + 3

Solve the number sentences and decide which answer is reasonable.

10 + 20 + 8 + 4 + 3 = 45

Write number sentences for each operation.

Since there are ten New York postcards, the probability of Si buying one of those is $\frac{10}{45}$, or $\frac{2}{9}$. After Si selects this postcard, there is one fewer postcard from which to choose. There are now 44 postcards left, eight of which are London postcards, so the probability of Si buying one of those is $\frac{8}{44}$, or $\frac{2}{11}$. To find the probability that he bought a New York postcard and a London postcard, multiply the two probabilities:

$\frac{2}{9} \times \frac{2}{11}$

Solve the number sentences and decide which answer is reasonable.

$\frac{2}{9} \times \frac{2}{11} = \frac{4}{99}$

The probability that Si bought a postcard of New York and a postcard of London is $\frac{4}{99}$. For more on this concept, see Chapter 9.

46. *Read the entire word problem.*

We are given the number of sides of a polygon.

Identify the question being asked.

We are looking for the sum of its interior angles.

Underline the keywords and words that indicate formulas.

The words *polygon* and *interior angles* indicate that we must use the formula for the sum of the interior angles of a polygon.

Cross out extra information and translate words into numbers.

There is no extra information in this problem.

List the possible operations.

The sum of the interior angles of a polygon is equal to $180(s - 2)$, where s is the number of sides of the polygon. We must subtract 2 from the number of sides of the polygon, and then multiply by 180.

Write number sentences for each operation.

$8 - 2$

Solve the number sentences and decide which answer is reasonable.

$8 - 2 = 6$

Write number sentences for each operation.

Now, multiply by 180:

$(180)(6)$

Solve the number sentences and decide which answer is reasonable.

$(180)(6) = 1,080°$

Check your work.

Since we used subtraction and multiplication to find our answer, we can check it by dividing and adding. Divide the total number of degrees by 180: $\frac{1,080}{180} = 6$. Now, add 2 to 6, and this should give us the number of sides of the polygon: $6 + 2 = 8$ sides. For more on this concept, see Chapter 11.

47. *Read the entire word problem.*

We are given the area and the height of a triangle.

Identify the question being asked.

We are looking for its base.

Underline the keywords and words that indicate formulas.

The keywords *base*, *area*, *height*, and *triangle* tell us that we need to use the formula for the area of a triangle.

Cross out extra information and translate words into numbers.

There is no extra information in this problem.

List the possible operations.

We are given the area and the height. We can rewrite the formula $A = \frac{1}{2}bh$ by dividing both sides of $\frac{1}{2}h$. The base, b, is equal to $\frac{2A}{h}$.

Write number sentences for each operation.

First, multiply the area by 2:

$2(1.98)$

Solve the number sentences and decide which answer is reasonable.

$2(1.98) = 3.96$ yards^2

Write number sentences for each operation.

Now, divide that by the height:

$$\frac{3.96}{1.2}$$

Solve the number sentences and decide which answer is reasonable.

$\frac{3.96}{1.2} = 3.3$ yards

The base of the triangle is 3.3 yards.

Check your work.

Since $A = \frac{1}{2}bh$, check that half the product of the base and the height equals the area, 1.98 square yards: $\frac{1}{2}(3.3)(1.2) = \frac{1}{2}(3.96) = 1.98$ yards^2.

For more on this concept, see Chapter 11.

48. *Read the entire word problem.*

We are given the measures of the hypotenuse and one leg of a triangle.

Identify the question being asked.

We are looking for the measure of the other leg.

Underline the keywords and words that indicate formulas.

We are told that we are working with a triangle, and the keyword *hypotenuse* tells us that we are working with a right triangle.

Cross out extra information and translate words into numbers.

There is no extra information in this problem.

List the possible operations.

Since we have a right triangle and the lengths of two sides, we can use the Pythagorean theorem to find the length of the missing side. Rewrite the formula $a^2 + b^2 = c^2$ as $b^2 = c^2 - a^2$, since we are looking for the length of a leg.

Write number sentences for each operation.

First, find the difference between the square of the hypotenuse and the square of one leg:

$(65)^2 - (25)^2$

Solve the number sentences and decide which answer is reasonable.

$(65)^2 - (25)^2 = 4,225 - 625 = 3,600$

Write number sentences for each operation.

The square of the leg is 3,600, so to find the length of the leg, we must take the square root of 3,600:

$\sqrt{3,600}$

Solve the number sentences and decide which answer is reasonable.

$\sqrt{3,600} = 60$ inches

The length of the other leg is 60 inches.

Check your work.

The square of the hypotenuse should equal the sum of the squares of the legs: $(25)^2 + (60)^2 = (65)^2$, $625 + 3{,}600 = 4{,}225$. For more on this concept, see Chapter 11.

49. *Read the entire word problem.*

We are given the area of a circle.

Identify the question being asked.

We are looking for a circumference.

Underline the keywords and words that indicate formulas.

The words *area* and *circumference* are keywords.

Cross out extra information and translate words into numbers.

There is no extra information in this problem.

List the possible operations.

The formula for area of a circle is $A = \pi r^2$, so we can find the radius by dividing the area by π, and then taking the square root of the quotient. Once we have the radius, we can multiply it by 2π to find the circumference.

Write number sentences for each operation.

Divide the area by π:

$\frac{169\pi}{\pi}$

Solve the number sentences and decide which answer is reasonable.

$\frac{169\pi}{\pi} = 169$ square feet

Write number sentences for each operation.

Find the radius by taking the square root of 169:

$\sqrt{169}$

Solve the number sentences and decide which answer is reasonable.

$\sqrt{169} = 13$ feet

The radius of the circle is 13 feet. To find the circumference, multiply the radius by 2π.

Write number sentences for each operation.

$(13)(2\pi)$

Solve the number sentences and decide which answer is reasonable.

$(13)(2\pi) = 26\pi$ feet

Check your work.

Divide the circumference by 2π to find the radius. Then, square the radius and multiply by π. This should give us back the area, 169π square feet. $\frac{26\pi}{2\pi} = 13$, $\pi(13)^2 = 169\pi$ square feet. For more on this concept, see Chapter 11.

50. *Read the entire word problem.*

We are given the volume of a sphere.

Identify the question being asked.

We are looking for its surface area.

Underline the keywords and words that indicate formulas.

The words *sphere*, *volume*, and *surface area* are keywords. We will use the formula for the volume of a sphere to find the radius of the sphere, and then use the formula for the surface area of a sphere.

Cross out extra information and translate words into numbers.

There is no extra information in this problem.

List the possible operations.

The formula for the volume of a sphere is $V = \frac{4}{3}\pi r^3$, where V is the volume of the sphere and r is the radius. We can find the radius of the sphere by multiplying the volume by $\frac{3}{4}$, dividing by π, and then taking the cube root. Since the formula for the surface area of a sphere is $SA = 4\pi r^2$, once we have the length of the radius, we can square it and multiply by 4π to find the surface area.

Write number sentences for each operation.

First, multiply the volume by $\frac{3}{4}$:

$(\frac{3}{4})(288\pi)$

Solve the number sentences and decide which answer is reasonable.

$(\frac{3}{4})(288\pi) = 216\pi$

Write number sentences for each operation.

Next, divide by π:

$\frac{(216\pi)}{(\pi)}$

Solve the number sentences and decide which answer is reasonable.

$\frac{(216\pi)}{(\pi)} = 216$

Write number sentences for each operation.

Now, take the cube root of 216 to find the radius:

$\sqrt[3]{216}$

Solve the number sentences and decide which answer is reasonable.

$\sqrt[3]{216} = 6$ centimeters

Write number sentences for each operation.

The surface area of a square is $4\pi r^2$, so square the radius and multiply by 4π.

$4\pi(6)^2$

Solve the number sentences and decide which answer is reasonable.

$4\pi(6)^2 = 4\pi(36) = 144\pi$ centimeters^2

Check your work.

Divide the surface area by 4π and take the square root. This will give us the radius of the sphere. Cube the radius and multiply by $\frac{4}{3}\pi$ to get back the volume of the sphere: $\frac{144\pi}{4\pi} = 36$, $\sqrt{36} = 6$; $\frac{4}{3}\pi 6^3 = \frac{4}{3}\pi(216) = 288\pi$ centimeters^3. For more on this concept, see Chapter 11.

Glossary

acute angle an angle that measures less than 90°

acute triangle a triangle with three angles that all measure less than 90°

addend each number being added in an addition sentence. In the number sentence 3 + 2 = 5, 3 and 2 are the addends.

adjacent angles two angles that are next to each other, sharing a vertex and a side

algebraic equation an algebraic expression with an equals sign, such as $3x = 9$

algebraic expression a single term or multiple terms on which one or one operations are performed and in which at least one variable is present

area the amount of two-dimensional space that a surface covers. Area is expressed in square units.

backward phrase a group of words and numbers that describe an operation in which the numbers are given in the opposite order that they will appear in a number sentence. For instance, *3 less than 4* is a backward phrase because when the numbers 3 and 4 are placed in a number sentence, 4 appears before 3: 4 − 3.

circle a set of connected points that are all the same distance from a single point, which is the center of the circle. A circle is a closed figure, but not a polygon.

circumference the distance around the outside of a circle

complementary angles two angles whose measures total 90°

constant a real number, such as 8 or –1, and not a variable

context the words that describe a situation or problem and give meaning to the words and number around them. For instance, the numbers three and five may not mean much by themselves, but in the context of a word problem, they could be added, subtracted, multiplied, or divided.

corresponding angles two or more congruent angles that have similar positions in a shape or figure

diameter the distance from one side of a circle to the other through the center of the circle

difference the result of subtracting one number from another. In the number sentence $5 - 3 = 2$, 2 is the difference.

dividend the number being divided in a division problem (the numerator of a fraction). For example, in the number sentence $4.5 \div 9 = 0.5$, 4.5 is the dividend.

divisor the number by which the dividend is divided in a division problem (the denominator of a fraction). For example, in the number sentence $4.5 \div 9 = 0.5$, 9 is the divisor.

equilateral triangle a triangle that has three congruent sides and three congruent angles

extraneous information details that are given in a word problem but are not needed to solve the word problem. Long descriptions, background information, or numbers that are not needed to find the answer to the word problem could be extraneous information.

fact family a group of related equations that use the same numbers. A fact family usually pairs addition and subtraction equations, or multiplication and division equations. The number sentences $5 - 3 = 2$, $5 - 2 = 3$, $2 + 3 = 5$, and $3 + 2 = 5$ are all members of the same fact family.

factor each number being multiplied in a multiplication problem. For example, in the number sentence $1.2 \times 6 = 7.2$, 1.2, and 6 are factors.

fraction a number that represents a part of a whole. A fraction itself is a division statement. $\frac{2}{8}$ is an example of a fraction. In this fraction, 2 is divided by 8.

greater than sign $>$, used when the first of two numbers is larger than the second of those two numbers. For example, *5 is greater than 4* can be represented as "$5 > 4$."

greatest common factor the largest number that divides evenly into two numbers. For example, the greatest common factor of 12 and 18 is 6. Both 12 and 18 can be divided evenly by 6. Other numbers (1, 2, and 3) are also factors of both 12 and 18, but 6 is the greatest common factor.

improper fraction a fraction that has a value greater than or equal to 1 or a value less than or equal to -1. $\frac{3}{2}$ and $-\frac{3}{2}$ are both improper fractions.

inequality a number sentence in which the quantity on the left side is greater than, greater than or equal to, less than, or less than or equal to the quantity on the right side of the equation. The number sentence $5x + 1 < 11$ is an inequality.

interior angle an angle found within a closed figure

inverse the opposite of an entity such as a number or operation. Addition and subtraction are inverse operations; 3 and -3 are additive inverses.

isosceles trapezoid a trapezoid with a pair of congruent sides, angles, and diagonals

isosceles triangle a triangle that has two congruent sides and two congruent angles

keyword a word that signals what operation to use. For instance, the keyword *sum* signals addition.

less than sign <, used when the first of two numbers is smaller than the second of those two numbers. For example, *4 is less than 5* can be represented as "4 < 5."

like fractions two or more fractions with the same denominator. For example, $\frac{1}{3}$ and $\frac{2}{3}$ are like fractions, because they both have a denominator of 3.

mean an average of a set of values. The mean is found by adding all of the values in a set together and then dividing by the number of values.

median the middle value of a set after values in the set are put in order from least to greatest. If there is an even number of values in a set, the median is the average of the two middle values.

minuend in a subtraction sentence, the number from which you are subtracting. In the number sentence 5 − 3 = 2, 5 is the minuend.

mixed number a number made up of a whole number and a fraction. The number $1\frac{1}{2}$ is a mixed number. Mixed numbers, like improper fractions, have a value that is greater than or equal to 1 or a value less than or equal to −1. The improper fraction $\frac{3}{2}$ is equal to the mixed number $1\frac{1}{2}$.

mode the value in the set that occurs most often. If there are two or more unique values that occur most often, then the set will have more than one mode.

obtuse angle an angle that measures greater than 90°

obtuse triangle a triangle with one angle that measures more than 90°

operand a number or variable used in an operation. In the number sentence 4 + 5 = 9, 4 and 5 are operands, and the operation is addition.

operation a procedure that is applied to one or more numbers or variables. Addition, subtraction, multiplication, and division are all operations.

operator the symbol for an operation. The symbols +, −, ×, and / are all operators.

opposite angles congruent angles that are formed by the intersection of two lines

parallelogram a quadrilateral whose opposite sides are parallel and congruent. Also, its opposite angles are congruent, its consecutive angles are supplementary, and its diagonals bisect each other.

perimeter the distance around an area. Perimeter is expressed in linear units.

polygon a closed figure made up of line segments

probability the likelihood that an event or events will occur, usually given as a fraction in which the numerator is the number of possibilities that allow for the event to occur and the denominator is the total number of possibilities.

product the result of multiplying two numbers. For example, in the number sentence $1.2 \times 6 = 7.2$, 7.2 is the product.

proper fraction a fraction that has a value less than 1 and greater than -1. For example, $\frac{2}{3}$ is a proper fraction, as is $-\frac{2}{3}$.

proportion a relationship between two equivalent ratios. The ratio 16:12 is equivalent to the ratio 4:3. These equivalent ratios can be expressed as a proportion: $\frac{16}{12} = \frac{4}{3}$.

Pythagorean theorem the sum of the squares of the bases of a right triangle is equal to the square of the hypotenuse of the right triangle: $a^2 + b^2 = c^2$.

radius the distance from the center of a circle to a point on the circle

range the difference between the smallest value and the greatest value in a set

ratio a comparison, or relationship, between two numbers. Ratios can be shown using a colon or as a fraction. The ratio *3 to 2* can be written as "3:2" or "$\frac{3}{2}$".

rectangle a parallelogram with four congruent, 90° angles and congruent diagonals

rhombus a parallelogram with four congruent sides and perpendicular diagonals that bisect their angles

right angle an angle that measures exactly 90°

right triangle a triangle with one angle that measures exactly 90°

quadrilateral a polygon with four sides

quotient the result of dividing one number by another. For example, in the number sentence $4.5 \div 9 = 0.5$, 0.5 is the quotient.

similar triangles triangles that have identical angles. All three corresponding angles are congruent, although the sides of the triangles may not be congruent.

square a quadrilateral with all of the properties of a rhombus and a rectangle: four congruent sides, four congruent, 90° angles, two pairs of parallel sides, consecutive angles that are supplementary, and diagonals that are congruent, perpendicular, bisect each other, and bisect their angles.

subtrahend in a subtraction sentence, the number that you are subtracting from the total. In the number sentence $5 - 3 = 2$, 3 is the subtrahend.

supplementary angles two angles whose measures total 180°

surface area the sum of the areas of each face, or side, of a solid figure

sum the result of adding two numbers. In the number sentence $3 + 2 = 5$, 5 is the sum.

term a variable, a number, or a variable and a number that is multiplied, divided, or raised to an exponent. Terms can be added or subtracted.

trapezoid a quadrilateral with at least one pair of parallel sides

triangle a three-sided polygon whose angles total 180°

unlike fractions two or more fractions with different denominators. For example, $\frac{1}{3}$ and $\frac{1}{4}$ are like unlike fractions, because they do not have the same denominator.

variable a letter or symbol in an equation or an expression that holds the place of a number

Venn diagram a drawing of one or more shapes that each represents a quantity of an item. The shapes can overlap to show items that are members of more than one set.

vertex the intersection of two rays or lines that form an angle

vertical angles congruent angles that are formed by the intersection of two lines

volume the amount of space taken up by a three-dimensional object. Volume is expressed in cubic units.